Security Metrics Management

Security Metrics Management

Measuring the Effectiveness and Efficiency of a Security Program

Second Edition

Dr. Gerald L. Kovacich

Edward P. Halibozek

AMSTERDAM • BOSTON • HEIDELBERG • LONDON
NEW YORK • OXFORD • PARIS • SAN DIEGO
SAN FRANCISCO • SINGAPORE • SYDNEY • TOKYO

Butterworth-Heinemann is an imprint of Elsevier

Butterworth-Heinemann is an imprint of Elsevier
The Boulevard, Langford Lane, Kidlington, Oxford OX5 1GB, United Kingdom
50 Hampshire Street, 5th Floor, Cambridge, MA 02139, United States

British Library Cataloguing-in-Publication Data
A catalogue record for this book is available from the British Library

Library of Congress Cataloging-in-Publication Data
A catalog record for this book is available from the Library of Congress

ISBN: 978-0-12-804453-7

For Information on all Butterworth-Heinemann publications
visit our website at https://www.elsevier.com

Publisher: Candice Janco
Acquisition Editor: Sara Scott
Editorial Project Manager: Hilary Carr
Production Project Manager: Punithavathy Govindaradjane
Cover Designer: Matthew Limbert

Typeset by MPS Limited, Chennai, India

Dedication

This book is dedicated to all the global security professionals who try to justify their assets protection programs, their budgets, and their jobs.

It is also dedicated to those who must protect assets from the insider and outsider threats but who now face the additional challenges of protecting those assets from global threats of terrorists, hackers, and other miscreants.

Contents

About the Authors

Dr. Gerald L. Kovacich has over 40 years of security, investigations, counterintelligence/counterespionage, drug-suppression, antiterrorism, information systems security, and information warfare experience in both the US government as a special agent and in business as a technologist and manager for numerous international corporations' security programs. He has also been a consultant to the US and foreign government agencies and global businesses. He has lectured internationally and has written numerous published articles and books on security. He conducts research and writes on such topics as global, nation-state, and corporate assets protection; economic and industrial espionage; netspionge; terrorism; and information warfare. He also is a writer of fiction to include poetry, short stories, and essays.

Edward P. Halibozek, MBA, has over 37 years of security experience. This includes experience in assets protection, government security, contingency planning, executive protection, and security management. He retired from Northrop Grumman Corporation where he was the Vice President of Security. He is the former Chairperson for the Aerospace Industries Association, Industrial Security Committee and, also served as a member of the National Industrial Security Program Policy Advisory Committee (NISPPAC). He is a lecturer on security and management issues and has written and published numerous assets protection plans, policies and procedures, and position papers related to corporate and government security. Currently, he is a part-time instructor for California State University, Fullerton. He also provides security consulting services in the health care industry and the communications industry. He holds a Master of Science degree in Criminal Justice from California State University and an MBA from Pepperdine University.

Other Books by Authors

OTHER BOOKS BY EDWARD P. HALIBOZEK AND DR. GERALD L. KOVACICH

The Manager's Handbook for Corporate Security: Establishing and Managing a Successful Assets Protection Program, April 2003, ISBN 0-7506-7487-3; published by Butterworth-Heinemann. (Pending Second Edition publication in Spring 2017).

Instructor's Manual for The Manager's Handbook for Corporate Security: Establishing and Managing a Successful Assets Protection Program, 2005, ISBN 13: 978-0-750-67938-1; ISBN 10: 0-750-67938-7, published by Butterworth-Heinemann.

Mergers & Acquisitions Security: Managing Security Issues Before, During and After a Merger or Acquisition, April 2005, ISBN 0-7506-7805-4; published by Butterworth-Heinemann.

The Corporate Security Professional's Handbook on Terrorism: Protect Your Employees and Other Assets Against Acts of Terrorism, 2008,
ISBN: 978-0-7506-8257-2; Butterworth-Heinemann; Kovacich.

BY DR. GERALD L. KOVACICH

Information Systems Security Officer's Guide: Establishing and Managing a Cyber Security Program, 2016, ISBN: 978-0-12-802190-3, Third Edition.

I-Way Robbery: Crime on the Internet, May 1999, ISBN 0-7506-7029-0; co-authored with William C. Boni; published by Butterworth-Heinemann; Japanese version published T. Aoyagi Office Ltd, Japan: February 2001, ISBN 4-89346-698-4.

High-Technology Crime Investigator's Handbook: Working in the Global Information Environment, September 1999, ISBN 0-7506-7086-X; co-authored with William C. Boni; published by Butterworth-Heinemann; Second Edition co-authored with Andy Jones; 2006.

Netspionage: The Global Threat to Information, September 2000, ISBN 0-7506-7257-9; co-authored with William C. Boni; published by Butterworth-Heinemann.

Information Assurance: Surviving in the Information Environment, September 2001, ISBN 1-85233-326-X; co-authored with Dr. Andrew J. C. Blyth; published by Springer-Verlag Ltd (London); Second Edition published in late 2005.

Global Information Warfare: The New Digital Battlefield, 2016, ISBN 13:978-1-4987-0325-3; co-authored with Andy Jones; published by CRC Press, Taylor & Francis Group, Second Edition.

Fighting Fraud: How to Establish and Manage an Anti-Fraud Program, 2008, ISBN: 13: 978-0-12-3708868-7; 10: 0-12-370868-0, published by Elsevier Inc.

Poems of Life: Thoughts of Human Experiences, 2012, ISBN: 978-1-4772-9634-9; 978-1-4772-9633-2; 978-1-4772-9632-5; AuthorHouse.

Book of Waking Dreams: Stories of the Dream Man, 2015, ISBN-10: 1504903706; ISBN-13: 978-1504903707; AuthorHouse.

Ramblings of an Old Man, 2015, ISBN-10: 1504909097, ISBN-13: 978-1504909099; AuthorHouse.

Essence of Her: Collected Poems, 2015, ISBN-10: 1504903684; ISBN-13: 978-1504903684; AuthorHouse.

BY EDWARD P. HALIBOZEK

Introduction to Security: Ninth Edition, 2013, ISBN:978-0-12-385057-7; co-authored with Robert J. Fischer and David C. Walters; published by Butterworth-Heinemann.

Foreword

When I wrote the foreword to the first edition of this book, some 10 years ago, I looked at where we had come from and the issues that were becoming apparent at that time. There has been significant change in the legal requirements, the threat environment, and the technologies that we are using since that time and it is wholly appropriate that this second edition has been produced. These days with cloud computing and the increasing use of virtualized environments, assets protection is no longer largely about the physical theft of assets. It is now much more about the theft of intellectual property (IP), as we have seen demonstrated by the creation of the term Advanced Persistent Threat and the escalating cost of cyber-attacks on organizations.

In the past, the people who were involved in security came mainly from one of two groups: law enforcement or the military; however, these days there is an increasing cadre of security professionals who have graduated from universities with relevant degrees.

As time has passed and the level of knowledge, experience, and business acumen needed to effectively protect assets has changed, there has been an increasing level of knowledge and a more comprehensive understanding of the need for a holistic approach to security. This has included the use of some of the very technologies that are part of the cause of the problems to assist in the protection of assets. This has not necessarily led to improved security, but has allowed organizations to keep pace with the developing and changing threats, however, the continuous reports that we see on security breaches and the scale of the losses that are being experience make it clear that there are still improvements and changes needed.

One of the major issues that still exists is that security, in many organizations is still not thought of as a business process and this should not be a surprise when a large number of the practitioners are science graduates or retired law enforcement or ex-military personnel who knew little of business principles and needs and do not speak the language of the business. Historically, security has been seen as a cost to the business (perhaps a hangover from the days of purely physical security) rather than as an integral part of business and a potential business enabler.

The second issue is that it is very difficult to measure any of the factors that are involved vis-à-vis costs versus defined benefits of security. Most organizations struggle massively with things such as the value of their assets, particularly when you look at things such as IP.

Since the first edition of this book, as the technologies have developed, we have seen increasing connectivity and globalization and organizations, both business and government, have increasingly connected to public networks and provided more access to their organizational network infrastructure.

In the global environment, lower production costs for products and services generally provide for a competitive edge. Since security is still usually viewed as an overhead cost, it is still often evaluated in terms of what must be done in order to achieve compliance or to meet the perceived requirement and not in terms of how security might add value to the enterprise. Even more than 10 years after the publication of the first edition, this still leads to security being viewed as an area that is a likely target for executives seeking cost reduction.

It is still true that all business is about risk and, depending on the risk appetite of an organization, it may be acceptable to accept higher levels of risks in order to reduce the costs of security. This is the prerogative of management but should be undertaken with a complete and accurate understanding of the cost versus benefit equation of the value of security and the implications of a failure.

In order to achieve the appropriate level of cost-effective security, the security expert who has come to understand the technology (or the "techie" who has gained an understanding of security) now also has to understand the business. It is also necessary for the management of the organization to understand and have some trust in the security processes.

Today, business is still very much the driving force in the use of information technology and computers and it is still struggling to integrate the way in which the security of the technologies can be dealt with in terms that are understood by the management. Security professionals have, in the past, managed this with expressions of doom and disaster if "security" is not taken seriously, but they have not, to date, been able to provide management with the detailed costs and benefits that the implementation of identified security measures would create.

In order for the business to be able to deal with security as it does with any other process, there is a need for security to be able to be expressed in the same type of business terms and to be able to have metrics applied to it that are meaningful to management, which will include details such as the costs and benefits.

At the time that the first edition was published, this has not been achieved with any regularity or consistency. The book attempted to move the integration of security into the business one step closer and this second edition brings this up to date. The book has been written by two people with a sound knowledge of, and experience in, both business and government security, as well as in the ways of measuring the effectiveness and efficiency of security. It has been written with not only the security professional in mind but also the non-security individuals who oversee security and have asset protection responsibility for the business or government agency. It is aimed at assisting both the security professional and management in dealing with the practical issues of the topic.

This book looks at the main aspects of assets protection (security) measurement and the types of metrics that can be applied to a range of security processes. It goes beyond the usual range that you might expect and extends to cover areas such as executive protection, contingency planning, and investigations. At all stages, the book addresses the value of metrics in helping the individual understand how effective and efficient security processes are. In essence it covers the cost versus benefit proposition on the application of security measures as part of an assets protection program.

The book was written for the benefit of all those involved in the business of security, from the practitioner to the board member, and done so in terms relevant to each of these audiences (i.e., measurement and the languages of business).

Prof. Andrew Jones, MBE PhD CITP CHFI CFIP MBCS M.Inst.ISP
Director, Cyber Security Centre, University of Hertfordshire,
Hatfield, United Kingdom

Preface

This book, *Security Metrics Management, Second Edition*, is an updated version of the First Edition while still keeping to its core philosophy, approach, and processes. It is still a primer on the topic.

There have been few—if any—new methodologies on how various aspects of how a business' and government agency's assets protection programs are measured. However, we believe that this timeless, *basic approach* we present in the First Edition and in this Second Edition is a good way to start measuring the effectiveness and efficiencies of a corporate assets protection program.

It is especially useful to a security professional new to the profession; as well as any others who have not measured the effectiveness and the efficiencies of their assets protection program or whose experience is limited.

To reiterate our approach: It is designed to provide *simple* and *basic* guidance to security professionals and managers and, students of security for establishing a baseline to begin the process of measuring the costs and benefits of their assets protection program—their security program—as well as its successes and failures—its effectiveness. It should begin where all assets protection policies, procedures, processes, plans, and projects should start—with the assets protection security corporate charter that identifies its drivers.

In other words:

- Why is assets protection needed at all?
- What drives that need?
- If needed, why are the related security functions needed?
- Even if they too are needed, why are they being performed the way they are being performed?
- Is the assets protection program working?
- At what costs?
- How are their costs measured?

- Can it be done more effectively (better)?
- Can it be done more efficiently (cheaper)?
- How?

This book will provide some methods to enable the reader to answer these questions in his/her environment. The book also includes a discussion of how to use security metrics to brief management, justify resources, and use trend analyses to develop a more efficient and effective assets protection program.

The security metrics management program (SMMP) that we discuss is not rocket science. It is a basic, rather simple, and hopefully commonsense approach to help you *begin* your process of managing a cost-effective assets protection program, some part of that program, or providing management oversight of such an assets protection program.

Once you have established an SMMP baseline, you should continue to improve on it and make it work for your specific environment. Furthermore, once this baseline is established, the measuring tools tested, refined, and implemented, one can then look at the availability of software applications that can help automate some or all of these measurement techniques.

> Like building a house, one must start with a strong, level foundation. If not, the house will not be properly built and may even fall apart. So it is with your SMMP.

COVERAGE

The intent of this book is to provide:

- a holistic approach to developing, implementing, and maintaining a basic SMMP;
- for the corporate or government agency security professional (as well as corporate management);
- an approach that will be useful to both the new and experienced professionals;
- for the security practitioner—and others such as auditors—methods which can enable them to measure the costs and benefits, as well as the success and failures of their security functions and overall assets protection program.

The approach used includes a fictitious international corporation called International Widget Corporation (IWC). This was done to help the reader look at an SMMP in a more realistic setting.

We emphasize that the information provided is generic and broad in scope in that it covers all major security functions for a basic international corporation. The methods and processes we offer can be applied by any security professional in any country. It will help provide an international answer to the problems of measuring security costs, benefits, successes, and failures across nations and societies.

Furthermore, we chose to organize this edition differently than the first edition. In this edition, we have placed all figures, charts, graphs, and surveys in the appendix thus allowing the reader to easily and quickly peruse through them together. For those readers used to seeing figures, charts, and graphs embedded within each chapter we hope you will find this approach useful, particularly as you seek to do comparisons among the various styles of figures, charts, and graphs.

NOTE

The reader is expected to be familiar with the basic functions and methodologies of assets protection. We addressed the corporate security profession in another book: *The Manager's Handbook for Corporate Security: Establishing and Managing a Successful Assets Protection Program*, April 2003, ISBN 0-7506-7487-3; published by Butterworth-Heinemann (Pending Second Edition publication in Spring 2017).

Dr. Gerald L. Kovacich and Edward P. Halibozek

Who Should Read This Book?

This book is for the new and experienced security professional, as well as those in government agencies, finance managers, and auditors who are involved with some aspects of understanding or managing security budgets and costs versus benefits. It is also useful for students of the security profession.

It is believed that the information provided in this book can be easily adapted by any security professional in any nation since an asset is an asset, so to speak, and measuring assets protection performance uses fundamental measurement techniques, analyses, graphic depictions, and project plans.

The use of security metrics management techniques is generic in nature and these techniques can also be used by executive management, auditors, and finance specialists to identify and track assets protection costs and performance. We attempt to use somewhat of a global perspective in writing this book so that it appeals to and can be used by security professionals and others around the world.

In addition, the following professionals will find this book useful:

- Corporate executives who have responsibility for protecting corporate assets as part of their inherent responsibilities to the corporation, its board of directors, and to shareholders. This includes executives such as the CEO, CFO, COO, CIO, CSO, and potentially many others. We will demonstrate how a security metrics management program (SMMP) can assist executive management in assessing the effectiveness and efficiency of their assets protection program.
- Corporate staff members who have specific responsibilities for specific assets: e.g., information technology staff, emergency services staff, and contingency/emergency planners.
- Professors, scholars, and researchers interested in the protection of people, information, and other assets within businesses or government agencies, as well as those who teach courses in business management, auditing, security, and criminal justice at colleges and universities.

In this book, we provide real-world examples of using an SMMP—e.g., case studies—of a security professional based on our own experiences and those of others that we know and show how security metrics management has helped support effective asset protection and management decisions.

When discussing the support agency—corporations, nonprofit agencies, government agencies, and the like—we will for the most part use the term "corporation," "business," or "company" to represent all of these entities. We used this approach to make it easier instead of each time detailing an entire list of entities. An SMMP is applicable to all these entities where security costs, benefits, successes, and failures should be assessed for effectiveness, efficiency, and the contribution to the enterprise.

CLOSING COMMENTS

We believe that this book, in the described format and with the identified topics, provides an exceptional security measurement foundation for security professionals or business/government agency executives who have a variety of levels of security experience and knowledge—in any location in any modern nation-state.

Our emphasis is not only on a global, modern-day world of business and government agencies, but also on providing sufficient guidance and tools that will support the inexperienced security professional in actually building a useful SMMP.

Please note that the numbers, flowcharts, and drivers are only provided as examples. We used a common, non-sophisticated (in our opinion) software product. A basic, simple product in support of a simple approach to baselining the costs and benefits of an assets protection program.

You will find this especially true with the data collections' numbers and their accompanying charts, which are fictional and are entered at random just to give you some samples. So, please concentrate on the methodology and the thought processes and do not try to concentrate on the accuracy of the numbers and whether the totals depict the cumulative numbers of whatever is being collected on a monthly or quarterly basis. Yes, we know that some of you out there will do that. So, one-plus-one may or may not show up as two. Again, (we stress) concentrate on the idea, process, formats, and methodology behind it all and not the actual numbers.

Some of the information provided on corporate security assets protection functions and related information is taken from our other published books. This was done so that we can provide a consistent and detailed background of information about corporate security functions that will help the reader understand our SMMP approach and philosophy.

Although not entirely redundant throughout each of the book's chapters, each chapter was written to be used as much as possible as an independent guide to one or more aspects of an assets protection program and its specific security functions. For example, if your focus is on guard forces and possibly other aspects of physical security and your security organization does not manage the information systems security function, you can go to the appropriate part of the book without reading about other security functions first (after reading the introductory chapters). We try to offer the security functions in a manner that allows you to pick and choose which ones you want to know first vis-à-vis security metrics.

You will also find that we are always (hopefully) making the point that using security metrics management can assist you in identifying and reducing the costs of an assets protection program and the related security functions. After all, security is generally an overhead cost to a corporation and therefore not only adversely impacts a corporation's potential profits but also has that "hidden" cost of lost employee productivity. Helping the security professional (and others) maintain an efficient and effective assets protection program is our goal.

As you read through this book, you will find that the SMMP is basically viewed as a combination of:

- Identification of drivers
- Documenting process flows
- Data collection
- Data analysis
- Graphically depicting the data to understand and tell the story

Over the years, we have found that graphically depicting data in chart format is a useful and necessary part of any SMMP. That is because, as the old saying goes, "A picture is worth a thousand words." This is certainly true when it comes to an SMMP. One can look at pages upon pages of data incorporated into spreadsheets and not be able to easily see the positive or negative trends.

Furthermore, from a management standpoint, the graphic depiction of data makes the results of the data collection easier to analyze, communicate, and generate corrective actions or make course corrections. Furthermore, executive management does not have time to view endless reams of data. Therefore, the charting of collected data is a good way to analyze data and brief management on anything from your assets protection program to why you need more budget, more resources, and everything in-between.

This book was developed primarily based on the knowledge and actual experience of the authors. Together the authors have decades of experience in developing and using SMMP. It has served well in all the international corporations where we have worked. We hope you have as much success as we have in using these techniques.

We thank you for reading it and would appreciate your comments, constructive criticism, and suggestions. Please send all comments to us through our publisher.

Dr. Gerald L. Kovacich and Edward P. Halibozek

Acknowledgments

Although we have also provided limited discussions on security metrics management in a few of our other books, the feedback we received from our readers and the attendees at our security metrics management-related lectures included requests for more details about developing and implementing a security metrics management program. We thank them for their support and for suggesting that we write a book providing more of a "how to," detailed approach to this topic. This book is written to fulfill those requests.

To the staff and project team of Elsevier—Sara Scott, Hilary Carr, and Punithavathy Govindaradjane. You all continue to be the best of the best!

To those other professionals in the Elsevier book publishing: we are grateful to all of them, not only for their support on this project, but also supporting our other projects over the years.

We also thank you and our many other readers for your comments, suggestions, and support over the years.

Introduction

It is important to understand what is meant by "security metrics management." So, before we get into the details of the matter, let us define and discuss some terms.

WHAT IS A METRIC?

To begin to understand how to use metrics to support management of a corporate assets protection program (CAPP), it is important to understand what is meant by "metrics." For our purposes, a metric is defined as a standard of measurement using quantitative, statistical, and/or mathematical analyses.

WHAT IS A SECURITY METRIC?

A security metric is the application of quantitative, statistical, and/or mathematical analyses to measuring security functional costs, benefits, successes, failures, trends, and workload—in other words, tracking the status of each security function in those terms.

There are two basic ways of tracking costs and benefits. One is by using metrics relative to the day-to-day, routine operations of each security function. Examples would be analyses of the costs of a security briefing program and conducting noncompliance inquiries (internal investigations into loss of assets). In more financial terms, these are the recurring costs.

Metrics can be and are used as individual data points. They are often best used in the depiction of trends. For example: Is the cost of security at company X going up or down?

The other way of tracking costs and benefits is through the formal project plans. Remember that security functions are "level-of-effort" (LOE), never-ending, daily work, while projects have a beginning and ending date with a specific objective and associated discrete costs.

So, in order to efficiently and effectively develop a security metrics management program (SMMP), it is important to establish that philosophy and way of doing business. Everything that a corporate security manager and security staff do can be identified as fitting into one of these two categories: routine operations (LOE) or projects.

In other words, project plans provide project schedules and are a tool to track time and expenses in relationship to the accomplishment of a task. Both time and costs (money) are metrics in and of themselves and are included in a project plan. A project plan establishes criteria and metrics (time, costs, milestones accomplished) that are used to measure performance to plan.

WHAT IS SECURITY METRICS MANAGEMENT?

Security metrics management is the managing of a CAPP and related security functions through the use of metrics. Security metrics management is the application of an individual metric or a set of metrics as a means of assessing the performance of a security process, security processes, or an entire security program. Through the use of metrics, the security cost versus benefit analysis becomes more quantitative and easier to understand and communicate in common business terms. Metrics help the security professional and others better understand the efficiency and effectiveness (value) of an assets protection program.

METRICS, MEASUREMENT AND MANAGEMENT

Some metrics are unique to their environment—e.g., software—while others can be ported to various environments. What we are offering here for the security professional is not some scientific, complicated, or "formal" methodology that requires training classes, nor years of experience to understand and efficiently and effectively use.

What we will be discussing throughout this book is a very basic and commonsense approach to begin to get a handle on the problem of identifying costs, benefits, success, and failures of assets protection programs and their related security functions.

Parts of that, as a Chief Security Officer, you already do—budgeting, for example. What we offer is an outline—an approach—that takes the methodology of metrics measurement and the philosophy of metrics management and combines them into an SMMP. In other words, putting it all together and using it as a security management tool to manage a CAPP.

Now that you have an understanding of what we mean by security metrics management, we can move on to an introduction to business and government agency security followed by detailed discussions of developing and implementing an SMMP as an integral part of an assets protection program and its related security functions, culminating into a look into the future of business security supported by metrics.

KEY WORDS AND PHRASES

The following key words and phrases should be understood by the reader and most certainly by the security professional:

1. Security Awareness
2. Physical Security
3. Personnel Security
4. Administrative Security
5. Security
6. Computer Security
7. Information Systems Security
8. Event Security
9. Information Warfare
10. Auditing
11. Compliance Assessments
12. Managing Assets Protection
13. Managing Security Organization
14. Assets Security
15. Assets Protection
16. Information Protection
17. Privacy
18. Liability
19. Risk Assessment
20. Risk Analyses
21. Cost–Benefits Analyses
22. Measurement
23. Metric
24. Processes
25. Process Improvement

CHAPTER 1

Security Metrics Management Program—An Overview

CONTENTS

INTRODUCTION

In the First Edition, we stated that: Some of the most common complaints of Chief Security Officers (CSOs) are that management does not support them. The other major complaint is that the costs and benefits of a corporate assets protection program (CAPP) are difficult to measure. More specifically, the effectiveness and efficiency of the program is difficult to measure.

Yes, those complaints are still heard echoing through the security departments; however, things are not as bad as they once were in that regard and yet in other ways, things are worse. With the advent of new, smaller, more powerful technology, as well as more and more wireless networking, more threats, vulnerabilities, and risks exist. Thousands of more and more attacks are taking place throughout the corporate world. Furthermore, they are more stealthy and more sophisticated while at the same time the simple theft of corporate property and simple, nontechnology-based frauds are taking place as in the past.

As for these complaints, the security professionals do get support because they are being paid—and these days more often than not, quite handsomely (although they might not think so)—and they have a budget that could have been part of corporate profits. Furthermore, respect is earned. If the security professionals wanted to be popular, they are definitely in the wrong profession. Yet, they may not be appreciated any more than in the past, but they are a "necessary evil" as corporate management may think of them.

One often hears executive management ask:

- What is all this security costing me?
- What am I getting for the money I spend?
- Is it working?
- How do you know it is working?

Security Metrics Management. DOI: http://dx.doi.org/10.1016/B978-0-12-804453-7.00001-X

- Can it be done at less cost?
- And of course, when some adverse event occurs: Why isn't it working? and What went wrong?

The more difficult question to answer is "What are the measurable benefits of a CAPP and of the security functions?" A security metrics management program (SMMP) can support the CSO in getting answers to these questions.

NOTE

Appendix has numerous samples of the charts, figures, etc. that we will be describing in this book.

FIRST STEPS IN THE DEVELOPMENT OF AN SMMP

The first step in the development of an SMMP to help answer your own and management's questions concerning security costs and benefits is to identify its uses. As we have discussed in the beginning, an SMMP can support security professionals in many ways. Some of the ways it can help are as follows:

- Identify the costs of an assets protection program.
- Analyze the security functional processes to identify individual functional costs.
- Identify areas where efficiency gains can be made.
- Identify the effectiveness of security functions.
- Track the costs and benefits of changes to security processes.
- Measure the successes of a CAPP.
- Demonstrate to management changes/improvements in the efficiency and effectiveness of security processes.
- Provide management with a performance assessment of the CAPP.

  Remember, it is imperative that, before developing an SMMP or in the first stages of its development, a series of actions should occur:
- Identify the security drivers—the reasons for having a CAPP or security functions which include, but are not limited, to legislation, regulations, contracts, organizational policies, and values.

  Depending on your corporation, you may have a "charter" that is the formal, corporate document that has been approved by corporate management and states why your organization is in existence. This then can be broken down into "drivers"—those things that are used to comply with your charter.
- Identify the statement of work (SOW)—what must be accomplished.
- Identify the processes used to accomplish the SOW—how does the work get done? Include the daily tasks performed by security

professionals (and sometimes others) to implement the CAPP or related security functions.
- Conduct a process analysis using tools such as process flow diagrams—these diagrams depict and document how each process works.

After the processes are documented in detail, the next step is to determine what is important about each process: What its purpose is or what it produces and then determine how it can be measured.

Each security function, CAPP policy, assets protection procedure, or process must have a purpose and that purpose should be directly tied to the security drivers. For example, what does the security function or process produce? What results are expected? For each process a measure must be established. That metric must reflect the purpose of the process. It must reflect what is being accomplished by that process. In total, the steps to accomplish this effort are the following:

- Identify each assets protection function/process.[1]
- Determine what drives that function/process, e.g., labor (number of people or hours used), policies, procedures, and systems: In essence, why does the function exist or why is the process used at all?
- Develop detailed security processes flow diagrams. They depict how the process works.
- Determine what gets measured.
- Establish a data collection process. The collection process may be as simple as filling out a log of transactions for later summarization and analyses. The use of a spreadsheet that can automatically incorporate assets protection actions/statistics into graphs is another method.
- Determine how the information should be depicted. For example, some data, to be useful, should be depicted in the context of trends. Other data should be depicted in terms of volume, cycle times, numbers of occurrence, or percentages. The tool used to meaningfully depict the metric will depend upon the metric itself and the message being communicated.

The decision to establish a process to collect data relative to a particular assets protection function or process should be decided by answering the following questions:

- What specific data will be collected?
- How will the data be collected?
- When will the data be collected?
- Who will collect the data?

[1]It is assumed each function costs time (e.g., labor hours), money, and equipment to perform.

- Where (at what point in the function's process) will the data be collected?
- What will the data depict?
- How will it be communicated?
- In what form will it be displayed?

By answering these questions for each proposed metrics chart, the CSO can better analyze the process as to whether or not a metrics collection process should be established for a particular security-related function. This thought-out process will be useful in helping explain it to the security staff or management, if necessary. It will also help the CSO decide whether or not the security staff should continue maintaining that particular metric collection process after a specific period of time.

The goal is to measure every process. Some processes more readily lend themselves to measurement than do others. All work should be measured eventually; however, perhaps only a critical few process metrics will be needed to determine the effectiveness and efficiency (cost vs benefit assessment) of the security function's process itself. Much depends on the environment in which you work—that is, your security duties and responsibilities.

> Building a good SMMP requires planning, project management, and continuous maintenance in order to constantly improve it over time.

Let us look at it from a CSO's actions. The CSO began with:

- an analysis of assets protection requirements (drivers);
- which led to identification of a CSO charter of duties and responsibilities;
- which led to the identification of assets protection functions and processes;
- which were graphically depicted in process flow diagrams.

Now that most of the work is done, the remaining task is to develop process-specific measures or metrics. As mentioned earlier, metrics should capture the essence of what the process is designed to accomplish. The metric should be meaningful in that it contributes to a better understanding by the CSO or process owner just how that process works and performs. Since needs and requirements change, all metrics should be reviewed, evaluated, and reconsidered for continuation on some periodic basis—annually, for instance—or when a requirement changes or the function must be changed. Keep in mind, as part of a quality management program, there must be the goal of "continuous process improvements."

Remember that, although the collection of data to develop robust and meaningful metrics for security processes will help the CSO better manage the assets protection duties and responsibilities, there is a cost incurred in the collection of data and the maintenance of measurement processes in terms of resources. These resources include:

- *People* who collect, input, process, print, and maintain the metrics for you;
- *Time* to collect, analyze, and disseminate the data;
- *Cost* of materiel, e.g., information systems hardware and software needed and used to support the effort.

One thing you must remember: The use of metrics is one tool to support many of the CSO's decisions and actions; however, it is not perfect. Therefore, the CSO must make some assumptions relative to the data to be collected. The CSO must remember that security metrics measurement and management is not rocket science, only a tool to help the CSO take better-informed actions and make better-informed decisions, as well as keep management abreast of the status of the CAPP.

It can also help you and executive management make better risk management decisions. However, as the CSO you should never get carried away with the hunt for "perfect data" or the "perfect metric," or become so involved in metrics data collection that "paralysis by analysis" takes place.

Keep in mind that tools used for metric collection and analysis such as spreadsheets, databases, and computer graphics programs can become very complicated with links to other spreadsheets and elaborate 3D graphic depictions. That may work for some, but the CSO should use the KISS (Keep It Simple Stupid) principle when collecting and maintaining metrics—at least to start. This is especially true if you, the CSO or security specialist, is just getting started and has no or little experience with metrics.

When choosing between automation and manual data collection, use what works the fastest to get started. Improvements can be made over time.

One may find that the project leads that are developing the "automated statistical collection" application are expending more hours developing the application that never seems to work just quite right, than it takes to manually collect and calculate the statistical information. When in doubt, go manual. You can always work on the automated collection processes over a longer period of time.

Remember, such projects are perfect for managing through the project management systems. This will keep the applications developers "on their toes" and not allow them to treat the project in an informal and lackadaisical manner.

Standardize all metrics charts to show a holistic, systematic, and organized SMMP.

It is also important from a managerial and "story-telling" viewpoint that all graphic depictions, charts, statistics, and spreadsheets be done in a standard format. This is necessary so that they can be ready at all times for reviews, analyses, and briefings to upper management or external customers. This standard is indicative of a professional organization and one that is operating as a focused team.

Those CSOs who are new to the CSO position, or to management in general, may think that this is somewhat ridiculous. After all, what difference does the format make as long as the information is as accurate as possible and provides the necessary details? This may be correct, but in the business environment, standards, consistency, and indications of teaming are always a concern of management. Your charts are indicative of those things.

It is hard enough for the CSO to obtain and maintain management support, so this should not be made more difficult than it has to be. Another negative impact concerning nonconformance of format is that the charts will be the topic of discussions by the attendees and not the information on them. Once the "nonconformance to briefing charts' standards" is discussed, management will have already formed a negative bias. From that point on, it will be more difficult to:

- get the point across,
- gain the decision desired,
- meet the established objective of the briefing.

It is better to just follow the corporate standards that are set than to argue their validity. It is better to save the energy for arguing for those things that are more important.

Of course the number, type, collection methods, etc., that the CSO will use will depend on the environment and the CSO's ability to cost-effectively collect and maintain the metrics. Find out and use the business-approved briefing chart formats, style, colors, etc. Such things as colors to be used are always important. For example, "green" often means good or satisfactory (denoting something positive) while "red" is used to identify unsatisfactory (negative), while "yellow" is used for caution.

Do not communicate a positive message using a negative color. Although the color yellow may be used to show caution, this color should be avoided unless it can show up very clearly on a chart. That is because such colors as yellow and lighter shades often do not clearly stand out on a chart. Undoubtedly someone in the audience will point that out to you, and this detracts from the point you are trying to make with the chart.

It is also important, when using line charts, to make the lines thick enough so that they are clear to the members of the audience in the back of the room. In fact, all charts should be developed based on the person in the worst position (usually the back of the room) being able to clearly see the information on the charts.

> If using text on charts, be sure they can be read by the members of the audience furthest from the briefing screen.

When using text on the charts, be sure that the text is also clearly visible. Do not use smaller fonts so that you can get all the required information on one chart. It is better to rework the chart and make two or more by subdividing the information and points you want to make.

One other point about charts: Photocopying charts in color may be cost-prohibitive and you always want to have copies of your presentation to give to each member of your audience (however, this is dictated by the size of the audience, purpose of the briefing, etc.). The copies of the charts as a handout to the audience may all be done in black and white. If that is the case, it is imperative that the information on each chart be as differentiated in black and white as it was by using colors. This may require you to design your charts with various contrasting color shades so that they can also be differentiated in black and white. For line charts, it is usually best to use a solid and then various dotted lines so that they also can be differentiated in black-and-white reproductions.

> If metrics charts are presented in color but hardcopies made in black and white, be sure that the information shown is just as clear and understandable.

As noted earlier, it also would not hurt to determine what executive management prefers in chart formats, colors, and so forth. This may seem ridiculous, but believe it. As trivial as it may seem for an executive of an international, multibillion dollar company to be concerned about chart formats and colors, it does happen.

Another way of stating this is to know your audience. Understand what your audience's expectations and preferences are as you build the presentation. Having a receptive and friendly audience at the onset of a briefing sets a positive and professional tone. Remember, each audience is not the same; they all have different expectations and preferences and it is incumbent on the briefer to learn this.

You may be wondering why we are spending so much time on this topic. The answer, if you have not discovered it so far, is that data collection usually must be converted into charts. As the old saying goes, "A picture is worth a thousand words." Besides, by using the charts to manage a security department and CAPP, you can easily show the trends, status, costs, and so forth of your efforts. Furthermore, it is a great tool for briefing executive management, your staff, visitors, and the like.

Make charts using the styles, colors, and formats that executive management prefers, to be sure you set the tone for them to concentrate on your charts' messages and not the charts themselves.

To add emphasis: Your secretary or administrative assistant is often in the best position to get the information on chart preferences and formats, as they often share the same tasks, albeit at a different bureaucratic level. Therefore, they may be in a better position to successfully work together on these issues while the CSO deals with other matters. *Note*: As a CSO, always be good to all the secretaries and administrative assistants in the business, as they are your best allies and hold the power associated with their bosses. They can be a valuable asset for you.

QUESTIONS CONCERNING DATA COLLECTION

Let us look at a method for determining security functional data (some may call it "statistics," but we will use the term "data") by asking and answering some basic questions concerning security functions as noted below. These are questions the CSO or project lead for an SMMP should be asking at the outset of an SMMP development, and periodically through the life of the SMMP to ensure it remains a reliable and useful tool.

- *Why should these data be collected?*
  One should be able to clearly state in one or two sentences why specific data is being collected.
- *What specific data will be collected?*
  The specific data that will be collected will be determined based on the security function and the specific need to track costs at specific

points in the process. It is also collected to determine other information that would be of interest to management and the CSO—and if it is of interest to management, it obviously better be of interest to the CSO.

- *How will these data be collected?*

  The data can be collected in a number of ways. They can be collected manually or through an automated data collection program. Generally speaking, it is usually more cost-effective to collect the data automatically; however, the development of an automated system may be too costly compared to a manual system. One must be cautious in that regard as the collection process should be done cost-effectively and efficiently. Again, remember KISS.
- *When will these data be collected?*

  When data are to be collected will be dependent on the need for the data. If a monthly report, annual report, quarterly report, or recurring briefing is to be accomplished, that may be the driver. However, as CSO managing a security department and CAPP, it would be logical to collect the data at the end of the month or the first week of each month, showing the previous month's data.
- *Who will collect these data?*

  The data should be collected, input, and maintained by the project leaders or process owners responsible for each security function. That way, they are part of the SMMP process and are also owners of their particular function. Furthermore, they should be responsible for leading the effort to identify more effective and efficient means of performing their security function(s).
- *Where (at what point in the security function's process) will these data be collected?*

The collection of data will be based on the need for the metrics and at the point where the data can be collected to fulfill that need. For some processes, it may be at the end of an individual security functional task, such as making a badge. Using that as an example, one would probably want to know how long it takes to process one badge for one new employee. After all, in today's environment, an employee identification badge is a critical tool and the employee "can't go to work" without a badge. Therefore, getting that person a badge the first day of employment may be mandatory.

One must remember that the longer it takes the new employee to process through the security badge office, the more it will cost the business in lost productivity for that new employee. The same would hold true when an employee must get a badge change or badge replacement.

SMMP CHART DESIGNS

We have alluded to the design of charts based on the data collected as part of the SMMP. However, this is such a crucial issue that we want to offer some "reinforcing guidance" on the topic.

The issue that will often come up when designing charts is what type of charts to use: bar, line, pie, etc. The choice should be to use the format that meets the chart's objective in the most concise and clear way, consistent with executive management briefing chart standards.

A CSO sometimes comes across numbers that are out of balance with each other—for example, 135 satisfactory security inspections ratings, 13 marginal. If the chart chosen were a line chart or a bar chart, the smaller number would be so dwarfed as to be almost unreadable. In this case, the pie chart may be the solution. The other solution could be to label each point in the chart with its number. For example, the bar or line designating 13 marginal ratings would have the number 13 over that point in the chart. This might give the perception that the marginal ratings were somewhat meaningless even though you, as a CSO, are aware that is not the case and that, in fact, it indicates increased vulnerabilities to successful attack on those assets. The pie chart, on the other hand, still shows that the number is small, but it at least appears larger on the graph. This allows the audience to look at that number as being more significant a matter than would be shown on the line or bar chart.

USING TECHNOLOGY TO DELIVER METRICS DATA

There are many reasons to deploy technology in the effort to develop, implement, and maintain an SMMP. Technology (computer-related hardware and software) can be used to improve the efficiency of how security services and products are delivered and this cost–benefits philosophy can be shown through security metrics charts. For example, you could make the case of costs–benefits through the charts to purchase new or upgraded computers and software, showing a return on investment graphically portrayed through metrics charts.

In today's tight resource environment, efficiency and timeliness are not the only motivation to seek alternative means of service and product delivery. The lack of budget to explore other options or add staff may drive you to seek help from technology. Furthermore, if you cannot get additional budget for additional resources that you have determined to be necessary to improve, then the use of technology to free already committed resources would help you provide assets protection as needed through security staff reallocation. In other words, this should allow you to redeploy those freed resources to other areas where they may be needed. An SMMP can provide the visibility needed to make the correct management decisions when looking at integrating more technology into security functions and the CAPP.

QUALITY AND OVERSIGHT

Remember that security is usually a cost center and not a revenue-producing entity; therefore, it is in competition with other organizations for budget. How that budget is obtained varies and is often dependent upon the demonstrated added value of the assets protection or security program to the corporation. As an example: If through an inspection it is determined that a security program is out of compliance and this noncompliant condition could adversely impact sales or a contractual obligation, management will divert resources to correct this problem. Yes, they will allocate money toward correcting the noncompliant condition, but you must prove your case. An SMMP helps you do just that.

The only concern you may have as the CSO is that someone other than you may be the person who receives the additional resources to fix the problem. This is of course assuming that the noncompliant condition was a result of something you did or did not do. It is not necessary to be completely dependent upon budgetary discretion of management. The solutions may rest within your own creativity as a CSO and willingness to explore other options. Furthermore, it is better to identify and correct security department and assets protection deficiencies before they are discovered by management. The SMMP can help you do just that. We suggest you take the initiative to do this on your own and not wait for a budget or compliance crisis.

It is a mixed blessing that security programs are often subject to very close oversight. It is good to have an external (outside of the security organization) perspective and assessment. But, dealing with these activities and organizations is time-consuming and can be difficult. Internal audit programs, government inspections, or customer security reviews are some of the various methods employed to determine the effectiveness of a security program. Dealing with audits and inspections takes much time and effort. However, they provide a CSO with periodic feedback on the condition of the security program. An SMMP can be developed and used as a quality and oversight support program. The other downside is that, more often than not, the focus is on compliance and not on efficiency. Efficiency needs to be measured and the good news is that the CSO can do this through an SMMP.

Perhaps the most important methods for measuring the efficiency and effectiveness of any security program are to have solid metrics in place, measuring all key processes for delivering products and services, and to conduct self-assessments (or self-inspections).

Conducting self-inspections allows a CSO to tailor the inspection or assessment process to focus on issues of efficiency and effectiveness. During a self-inspection or self-assessment, special emphasis can be placed on assessing efficiency or effectiveness of processes and the delivery of products and services. This in turn can be documented by the SMMP.

SECURITY METRICS AND PROCESSES

Because processes are such an integral part of an SMMP, let us look at them in more detail. Remember that basically a process is a series of steps or actions that produce something. In our case, the steps taken help protect assets.

When developing processes, a process statement should be developed that describes the actions to be taken and not the job or task. For example, "Produce a physical access violation report" and not "Physical Access Violation Report." In other words, there is action, using a verb instead of a noun.

Upon completion of that task, you must then decide how to produce the right metric or metrics. In order to do so, you must ask yourself:

- First: Are you measuring the product or the process which produces it?
  - An example would be looking at an employee climate-sensing product and process relating to their support for an asset protection program.
  - If you are tracking the climate itself (employees more or less are supporting the CAPP), you are measuring the product.
  - If you are tracking how well you obtain that climate data (e.g., How long does it take to get it? How accurate is the data?), you are measuring the process.
- Next, select the proper product or process metric, so as to measure its performance to target or customer specification; for example:
  - Product
    - Metric: % of reports delivered on time
    - Measurement: 98 of 100 reports were delivered on time
    - Result: 98% of reports were delivered on time

- Process:
  - Measure a key characteristic of the product or the process itself (often cost, quality, cycle time, or customer satisfaction).
  - Look for trends over time.
  - Make sure it is a key characteristic the process has input to or control over. For example:
    - Measuring the number of alarms a guard force responds to that are false alarms—they do not cause or control the alarm.
      - Measuring how quickly they respond or the percentage that are actual alarms that they respond to in less than 5 min—they do control that.

COST-AVOIDANCE METRICS

As a CSO, you may want to use the SMMP approach to be able to quantify the savings of some of your decisions. For example, when analyzing your budget and expenditures, you note that travel costs for your staff is a major budget item. This is logical since staff, as well as you, must travel to the various corporate offices to conduct assets protection inspections or assessments.

Again, using the project management approach, you lead a project team of yourself, staff members, and a representative from the Contract Office and Travel Office. Your goal is to find ways to cut travel costs while still meeting all CAPP objectives and fulfill your charter responsibilities. In other words, you are trying to avoid costs.

USING METRICS CHARTS FOR MANAGEMENT BRIEFINGS

The data collected must be shown in some meaningful way. The use of security metrics management charts is just such a way. When using these metrics charts for management briefings, remember the purpose of the briefing. A succinct message is being communicated to management. This is most effectively accomplished with graphic depictions of data in a meaningful way.

How that data is depicted will depend upon the message being communicated and the type of metric being used. For example, if a cost of security trend over a 5-year period is being depicted, the use of a bar chart suits the metric well. Or would a line chart be better? You decide based on your environment and briefing standards.

When the best form of depicting the metric is not clear, the CSO should experiment with various types of graphic depictions, as the use of line, bar, scatter, and pie charts can be very useful in delivering a point. It is recommended that the charts be kept simple and easy to understand. Remember, the old saying, "A picture is worth a thousand words." The charts should need very little verbal explanation. The message must be kept clear and concise. Do not get bogged down in details which detract from the objective of the chart's message.

One way to determine if the message the charts are trying to portray is clear is to have someone unfamiliar with the subject look at them and determine if the data is easy to understand. A CSO may choose to have security staff members look at the charts and describe what the chart tells them. If it is what the chart is supposed to portray, then no changes are needed. If not, the CSO should then ask the viewer what the chart does seem to represent and what leads them to that conclusion.

The CSO must then go back to the chart and rework it until the message is clear and it is exactly what the CSO wants the chart to show. Each chart should have only one specific objective and the CSO should be able to state that objective in one sentence; for example, "This chart's objective is to show that assets protection support to the business is being maintained without additional budget although the workload has increased 13%." In such a chart, emphasis is on showing that the workload has increased, budget has not increased, but assets are being adequately protected. That view should be perceived with little or no additional commentary.

SEQUENCE OF SMMP CHARTS FOR MANAGEMENT BRIEFINGS

When establishing a briefing to management in which the metrics charts will be used, it is recommended that a sequence of charts be developed. Think of it as telling a story. It has a beginning, middle, and an end. In this case, the story of security or a security function or issue is being told. It may even include the "moral" of the story—not in the true sense but in the context of something learned from the effort. The sequence may look something like the following:

- Assets protection drivers
- Specific drivers showing relationship with each specific security function
- Flowchart of each security function
- Flowcharts of each security subfunction using a work-breakdown type of approach
- The metrics charts for each part of a flowchart where time (productivity) and costs can, and should, be measured

When developing an SMMP, it is important to first outline the SMMP starting with an overall outline, which then includes details down to each flowchart of each security function. This process can be done by viewing it through the security functional view of the security department or through the view of the CAPP, the difference being that the security functional view only deals with security activities whereas the CAPP view deals with assets protection from an overall business perspective—what business employees must also do. Another way of looking at this is to start with the micro (a single security process or function) and move to the macro (the entire CAPP or maybe the entire security department or both). You may instead want to start with the overall CAPP view and gradually move to a more micro view.

The approach you use will be dependent on your philosophy as a CSO or your executive management needs, or how management wants to see such things. One way to find out how management generally likes to have the briefings done is to talk to those who prepare executive management briefings and also the executive management administrative staff. As we mentioned earlier, the secretary and other staff members have a clearer view of what their bosses would like to see than you would.

Remember that if executive management has some quirks about briefing charts, you should find out about it first so that your briefing is well-received instead of generating complaints about the types of charts or colors used. This has happened more than once, and in the end, the objective of the briefing was lost!

It is important that when developing an SMMP that you use a corporate-approved project plan format and corporate standard project planning software. That is important because when briefing corporate management on the status of assets protection, you want to also introduce them to your efforts to try to provide balanced assets protection with costs and acceptable risks.

That project plan should include the following:

1. SMMP objective: Establish a program to track costs, benefits, successes, and failures of a CAPP and related security functions.
2. Identify assets protection drivers—develop chart of all drivers.
3. Identify security processes related to each driver—develop charts connecting drivers to security processes.
4. Identify assets protection policies—develop charts to relate each major policy to the drivers and security functions.
5. Identify the security processes flowcharts used to support each major policy and procedure.

6. Develop individual high-level security process charts, one for each process showing level of effort (LOE), impact, costs, and benefits.
7. Develop lower level charts for each security subprocess as in #6, above.
8. Chart each major aspect of the CAPP to drivers and security processes.
9. Develop total budgetary chart and subcharts based on budget breakdown.

It is recommended that, to begin this process, the outline should concentrate on an SMMP from the view of the CSO looking inward at the security department's security processes and functions. This reason for this is that it addresses most of what is needed when viewing the CAPP but also would be a higher priority view from the CSO's point of view. Why? Because the CSO will be spending most of the time managing the department in a macro sense, driving down into the micro when problem-solving or change is necessary.

Of course, as the CSO, you would want to get more specific and track to a more detailed level of granularity than may be required for executive management briefings. In fact, the security staff responsible to lead a specific function should be tasked with developing the charts relative to their function, based on standard formats or boilerplates. That way, the staff knows exactly why they are doing what they are doing.

The next step would be for them to track their workflow and analyze it. At the same time, they would also look at current costs and cost savings, with the goal of identifying more efficient ways to successfully accomplish their jobs.

The benefit of such an approach is to get the security staff involved so that they feel ownership with their security function and therefore will be in a position to help identify steps to improve the processes, as they are closest to the function and processes.

Data collection costs time and therefore money—use both wisely.

SMMP AND EXECUTIVE MANAGEMENT

Remember that the data collected costs time and of course "time is money." Therefore, it is again emphasized that the data collected must be only that data that is needed and not "nice-to-have" data, which is a waste of time and therefore money. Also, no data should be collected for "just in case we need it later." However, this is a CSO or management decision which may be based on management's track record of asking "unexpected" questions for which the CSO has been more than once caught off guard.

Remember that a CSO is also an assets protection "salesperson" and must effectively advertise and market assets protection to management and the business' employees. So, the SMMP can assist in:

- justifying the need for more budget and other resources;
- indicating that the CAPP is operating more efficiently;
- helping to justify why budget and other resources should not be decreased.

  When deciding to develop metrics charts to track workload, efficiency, costs, or other measures of a function, always start at the high level and then develop charts in lower levels (more details) that support the overall chart. This is done for several purposes:
- The CSO may have limited time to brief a specific audience and if it is an executive management briefing, the time will be shorter since busy executives have much responsibility and little available time. They often focus on the "bottom line" or the essence of an issue.
- The "top–down" approach will probably work best. If you have time to brief in more detail, the charts will be available.
- If executive management has a question relative to some level of detail, then the other charts can be used to support the CSO statements and/or position in reply to the majority of the audience's questions.
- As part of the CSO's use of the SMMP, the top–down approach also provides flexibility for the CSO when using a systems approach to the data analyses—looking for costs, negative and positive trends in various levels of details, for example.

USING METRICS TO DETERMINE SUCCESS

The overall security process rating can be determined by a roll-up of the performance-to-target/product metric for the key processes. The formula is as follows:

1. Establish the percentage of success of each key process:

   Score/100 = percentage to success (e.g., the goal for Security Awareness Training was 100%. During this particular rating period, we achieved this 95% of the time. 95/100 = 95%. On the other hand, if you have a numerical score as in Access Control, set an arbitrary rating scale; the goal is 0 occurrences of unauthorized access, but you might set a scale whereby 0 incidents = a score of 100, 1 = a score of 99, etc.).

2. Average the results of the key processes; this gives you the overall security rating. To convert this to the standard stoplight chart, use the following scale:
   - 100–90 = green,
   - 89–80 = yellow,
   - < 80 = red,

   with one exception: any score of less than 100 on certain critical key processes (e.g., not responding to the Research Center alarms in less than 5 min) requires a rating of "yellow" no matter how high the overall numerical rating is. Two consecutive months of less than 100 in those critical key processes require an automatic red, as does more than two incidents in a quarter or four within a year.

You may wish to change these breakpoints to reflect realistic performance or customer sensitivity to certain processes in your environment.

SUMMARY

Metrics management techniques will provide a process for the CSO to support assets protection and other security-related decisions. The CSO should understand that:

- Metrics management is an excellent method to track security processes and functions related to LOE, projects, effectiveness, efficiency, costs, proper deployment of resources, etc.
- It provides a methodology for measuring the success of a CAPP.
- The information can be analyzed and results of the analyses used to:
  - identify areas where efficiency improvements are necessary;
  - determine effectiveness of security functional goals;
  - provide input for performance reviews of the assets protection staff (a more objective approach than subjective performance reviews of today's CSO's);
  - where security services and support to CSO requires improvement, to meets its goals, etc.

Building a successful SMMP requires the use of a formal process to validate each step. Once that is completed for each security function, it can be summarized as a sort of abstract of that security function.

CHAPTER 2

Corporate Assets Protection Program

CONTENTS

INTRODUCTION

The corporate assets protection program (CAPP) is the primary vehicle used to incorporate all the security charter, that is: stating why security exists and what it is suppose to do. The role of security is a protective role. Protection of people, information, and physical assets is the leadership purview of the security professional and the security organization.

Historically, security has been viewed as a necessary support function. In many corporations, and certainly in government, this is changing. More often we are seeing security characterized as a mission-essential function and not just a support function.

How and why a corporation protects its assets are institutionalized in a CAPP. The CAPP can be viewed and managed as the overall continuous plan to protect corporate assets or it can be viewed as a program. The choice is based on the corporate culture or how the Chief Security Officer (CSO) decides it is best to manage the effort. Within International Widget Corporation (IWC), the CAPP is a program and also a macro security plan for protecting corporate assets. Approaching the CAPP as a plan offers the advantage of dividing that plan into subplans or logical (separations) sections that deal with the different aspects of the assets protection effort.

The method used may or may not have a bearing on how security metrics is used to manage the assets protection efforts. In any case, the CSO must first identify exactly what is needed as far as data to manage the entire program.

As a CSO, what data do you think you need? For example:

- To be able to manage the assets protection program?
- What data will assist you in managing the program effectively?

Security Metrics Management. DOI: http://dx.doi.org/10.1016/B978-0-12-804453-7.00002-1

The answer is not the same for all corporations or security departments. Each corporation is different and operates in an environment unique to them. Furthermore, what is most important to one CSO may not be as important to another. With that in mind, each CSO must develop and use data and security metrics measurements that work best in their own environment.

As a new IWC CSO, one would first want a macro view of the organization before examining the more micro processes. In reality, that will occur soon enough as the CSO is presented with the daily problems of any security operation. So, what are some of the more important security metrics management program (SMMP) data that would be needed?

Let us start with the CAPP's basic objective or goal: The CAPP is a tool to protect corporate assets and mitigate risks to the corporation. Ideally, the goal is to accomplish this objective in as effective and efficient a manner as possible. Therefore, any data collection through security metrics management charting and oversight should be focused on those goals:

- Is the CAPP working as planned?
- Is it effective?
- How much does it cost?
- How can it be done better?
- How can it be done cheaper?

A CSO would want to begin with a security overview chart followed by the functional charts used in earlier chapters, in which the security drivers are identified. This will help refresh the CSO and others as to what drives the need for a CAPP. There are various ways to view these drivers and to flowchart them.

THE CAPP AND OTHER DRIVERS, PLANS, AND THEIR FLOWCHARTS

The CAPP must be (or at least should be) integrated, to include incorporating the other security plans such as the annual, tactical (short-term), and strategic (long-term) plans. These plans in turn must be integrated into the overall corporate plans. After all, the security department is a service and support organization that does perform mission-critical tasks.

Security plans should be integrated into and support corporate business plans.

As a CSO, you should develop such flowcharts, of course in more detail, that demonstrate how security goals and objectives (and other security activities) flow into the corporate business plans. A graphic depiction of how security processes integrate into corporate business plans is a useful tool when demonstrating the value security brings to the corporation. After all, if you are not supporting the corporate plans, you are marching to the wrong drummer.

CAPP DATA COLLECTION AND SECURITY METRICS MANAGEMENT

As a CSO, you want to be able to pictorially show the relationship between the security department's goals, objectives, projects, functions, plans, policies, procedures, processes and how they relate to providing service and support to the corporation. In today's corporate world, you must look and act as a "team player." Such flowcharts and security metrics charts assist in that portrayal.

Once you have developed such flowcharts, "connecting the dots," you must decide on what security metrics you want to collect, analyze, and use as part of your SMMP, overall CAPP, and security department management functions.

Earlier in this chapter, we discussed key questions that you as the CSO must be able to answer and explain to the corporation's senior management, since they need and want the answers too. If you recall, these questions are the following:

- Is the CAPP working as planned?
- Is it effective?
- How much does it cost?
- How can it be done better?
- How can it be done cheaper?

Let us discuss these one at a time from the SMMP perspective.

IS THE CAPP WORKING AS PLANNED?

The goal of the CAPP is the protection of information, people, and physical assets. It is a plan that is documented and contains goals and objectives, as well as the details of the assets protection program. It must be documented to ensure its availability to those who need it and to use as a baseline for measuring against performance.

In order to determine if the CAPP is working, many measurements will be needed. Ideally, the fewer measures the better, particularly when attempting to depict success or failure to senior management. One of the ideal depictions would show the relationship of the value of corporate assets to the actual losses experienced over a defined period of time.

The overall chart would show the amount of corporate assets in one form or another, broken down into values for people, information, and physical (facilities and equipment) assets. However, when it comes to the value of some of these assets, that may be difficult to assess (e.g., what is the cost of each piece of information? what is the value to the company of an employee's life?).

That would not be an easy task or maybe not even practical. However, we can collect data from the investigative organization as to the losses of or damage to assets. Also, we can collect information from other IWC organizations, such as health and safety personnel or property management. What we can more readily collect is information showing where the CAPP is not working from the standpoint of protecting corporate assets.

In addition to the overall data collection and their related metrics charts, other data should be collected and their related "backup" metrics charts would be developed based on data collected showing:

- Categories of asset losses
- Their value (in current terms of the cost of replacement)
- Time-productivity lost
- Vulnerability that allowed the loss or damage to occur
- Consequences of that loss or damage
- Corrective action to be taken
- Cost of the corrective action versus the cost of the assets (e.g., if the corrective action costs more than the estimated losses or damages over a period of time, say 5 years, then maybe it is not worth correcting the vulnerability). After all, it is a matter of costs–benefits.
- If corrective action is deemed appropriate, a project plan for that action

NOTE

Of course this can also be depicted by year or even over a 5-year period, as trend charts are valuable charts to manage assets protection. If the trend is positive (downward loss trend) then as a CSO, you should find out why and praise/reward those responsible. After all, they certainly catch hell when something goes wrong! So, why not also recognize them when something goes right? If the trend is negative, identify the reasons and establish projects to correct the problems.

IS IT EFFECTIVE?

As a CSO, how will you determine if the CAPP is effective? In other words, are all the parts that make up the IWC CAPP designed and integrated to such an extent that the entire IWC assets protection effort is effectively working?

As the CSO, you can get a sense of whether or not it is effective by looking at the trend charts of damages and losses of assets over time. If they are low or the trend is neutral or indicates less damages or losses, then it is effective.

By the way, when discussing losses, we mean all losses that include loss due to theft, damage, misappropriation, and "I don't know what happened to it but it must be here somewhere—I think." (Assuming it cannot be found in a reasonable amount of time.)

One of the important data points that ought to be depicted in one of your "subcharts" is one showing recoveries and the value of the recovered assets with and without the cost of recovery.

Based on the results of the data collected and the trend charts supported by the data, as a CSO, you would implement projects to identify and mitigate losses using a risk management approach.

HOW MUCH DOES IT COST?

The costs of the CAPP would be measured by the total costs of the security department—that is, the security department's total budget, as well as factoring in the sum total of all costs as collected through each security function where the productivity impact to nonsecurity personnel has been calculated—and that should be identified by each security function.

The process for data collection for this metric would be through identifying the cost data for each security function summed plus the security department's budget, excluding any redundancies, of course.

Remember, if assets did not need protecting, there would be no need for a security department. Therefore, the security department's budget as a minimum must be included in your cost calculations. It is the easiest data to obtain—assuming you as the CSO have a total security budget. If not, well, that scary thought is beyond addressing here. However, one "war story" may be worthy of telling here:

> One of us had once worked in a very large international corporate division. When hired and asked about the organizational budget, the boss, a senior executive and vice president, said that there wasn't a budget but we were all expected to spend only the funds necessary to meet the needs of the organization and successfully accomplish our objectives. Well, you can

guess the results. Sure enough, there were cost overruns on some contracts and most of the business unit's managers were spending as if they were multibillionaires without a money care in the world—clearly a poor planning process that ended up being very costly to the corporation.

This author took it seriously and, as usually happens, when the budget crash hit, we were all affected. So, while others already had many of their desired wish list items, we in security went wanting and also unrewarded for our conservative and ethical efforts. So, one word of advice (and as a security professional, one should never expect rewards anyway), let your ethical conscience be your guide if you are ever placed in that position!

Think about what back-up, detailed information (in data collection and metrics charts form) should be available to help you answer some obvious questions. For example, why the increase or decrease in specific years? If you were using an SMMP to manage the CAPP over the years, you would know that answer.

As a CSO, you may want to change the metrics chart and use a different one which explains why the upward or downward trend has occurred. It also may be a good idea to do an annual average chart over the 6-year period to show the long-term trend. In other words, develop your data collection and subsequent metrics charts for analysis, decision making, and briefings using a work-breakdown process.

HOW CAN IT BE DONE BETTER?

Doing it better—more effectively—is a product of taking what was learned from the data collection and metrics charting of an SMMP and using that information to improve related security processes. Tracking that performance can be accomplished by comparing the year-to-year trends in losses as a percent of asset value. Ultimately, changes should be made to the point where there are diminishing returns. That is to say, it costs more in security expenses than in value of losses prevented.

HOW CAN IT BE DONE CHEAPER?

Doing it cheaper—more efficiently—usually involves activities such as process analysis (including time-motion studies where appropriate), process modifications, process improvement, and quality management techniques. (There are many others but those stated tend to fit well with the analysis of security processes and functions.)

These management tools actually do work when properly applied but often get a "bum rap" by those who fight change or do not know how to properly apply these types of tools. All too often they are not properly used, or employees are not properly trained as to how to use these tools, which in turn causes the employees to discredit the processes.

One often hears the employees' cries, "Why can't we just do it like we've always done it?" What would be your reply? Obviously, as the CSO and in a leadership role, you should explain the corporate and business environment, the global marketplace issues, competitive advantage philosophy, and the goal of making each security function and its related process as effective and efficient as possible.

The security staff should also be told that this is not a way to downsize the security staff (although it certainly may cause that to happen—one never knows at the outset what the results will be). As is usually the case, one can always find work for the security staff and one approach may be to only eliminate positions when required to do so by upper management.

CASE STUDY

The IWC CEO is concerned about the cost of the CAPP and wants to know what exactly is involved in protecting the IWC assets. As the CSO, how would you respond?

Since the CEO's time is limited, the CSO decided to provide a 10-min briefing using no more than five charts with each chart to be briefed for no longer than 2 min. Ten minutes are also set aside for comments and questions, for a total of 20 min.

Which type of metric, in what type of graphic depiction, if any in this chapter, would you choose to use?

- Those which show assets expenditures, losses, and damages with their associated costs
- Ratio of costs to damages and losses
- Those showing efficiency gains over the last 6 years or whatever years are available
- Any data and related metrics charts showing a favorable return on investments

The data and metrics charts used may of course be different depending on the CSO, the CSO's SMMP, CEO's request, and the CEO's preference for briefings, as well as the corporate culture and environment.

SUMMARY

The CAPP is a program consisting of the following:

- List of corporate assets protection drivers (reasons for security and protection).
- Policies, procedures, processes, and plans used to define protection parameters, requirements, projects, and objectives.
- It is institutionalized in document form and available for all who support the CAPP for review and consultation.

The security metrics management data collection and graphics depictions are tools used by the CSO to help manage the CAPP. Data is generally collected, analyzed, and depicted by function and as a total organization.

The key to the security metrics of the CAPP from a management standpoint is the ability to answer, through data and supporting metrics charts, the following questions:

- Is the CAPP working as planned?
- Is it effective?
- How much does it cost?
- How can it be done better?
- How can it be done cheaper?

CHAPTER 3

Personnel Security

CONTENTS

INTRODUCTION

The personnel security function at International Widget Corporation (IWC) has two primary responsibilities:

1. Pre-employment and background investigations
2. Workplace violence (WPV) prevention program support

These are both vital functions that are an integral part of any corporate assets protection program (CAPP). You may not agree with some of the criteria used to screen out potential employees—such as convicted felons who have served time and paid their debt to society, or someone who had filed for, or still is in, bankruptcy. However, establishing the criteria is the prerogative of the company, in this case IWC, and must be done in accordance with all applicable laws and regulations, company policy, and values. IWC must establish its own threshold for risk in regards to the suitability of employees.

Remember that the screening process is done to try to hire only the most qualified, honest, and capable employees. Only the most suitable for IWC should be hired. Employees must be suitable for the work environment and they must be trustworthy. After all, much will be expected of any IWC employee and they must be capable of delivering. That delivery includes the ability and willingness to protect corporate assets.

At IWC, this responsibility is shared with the Human Resources (HR) staff who are responsible for the hiring process. Security is in a support role.

PRE-EMPLOYMENT AND BACKGROUND INVESTIGATIONS

Every company wants to employ competent and trustworthy people and IWC is no exception. IWC cannot afford to have employees who are incompetent or untrustworthy. This is a recipe for problems at best and disaster at

Security Metrics Management. DOI: http://dx.doi.org/10.1016/B978-0-12-804453-7.00003-3

worst. Incompetent employees are not capable of producing the best products and services.

Screening out the potentially bad employee prior to hiring is a much more effective and painless process than trying to get rid of an employee who has just been determined to be untrustworthy.

PRE-EMPLOYMENT AND BACKGROUND INVESTIGATIONS DRIVERS AND FLOWCHARTS

As with each security function, the security drivers should first be identified. Why have the functions at all? If it is found to be a necessary function, then the next question should be, "Why do it the way we are doing it?" And please do not ever accept an answer from anyone who says, "We've always done it this way." If you hear that reply, you have almost a guaranteed certainty that no one has ever looked into the process in any detail to see if it is being performed as effectively and efficiently as possible.

The overall security driver is of course the need to protect the assets of the IWC owners—the stockholders—as required by laws, regulations, and owner expectations. By screening potential employees and hiring only the most suitable persons, the risk to IWC assets should be decreased—at least that is the theory. However, neither process nor person is perfect, but one does the best one can. It is much better than not screening potential employees.

If you are the Chief Security Officer (CSO) or other security professional responsible for a security organization or function, regardless of how long you have been in the security business, never, and we mean never, accept that a particular function must be performed, unless it is backed up by valid security drivers. Once that is established, the next task for the CSO is to do what?

If you answered "Develop a personnel security functional flowchart," you are absolutely—wrong. The next step is to establish a project and a project team with the objective of determining how best to perform this function in the most effective and efficient way.

Of course, the project "team" may consist of primarily one person; however, even that one person will need assistance from time to time. Once the project is activated and the security drivers specific to that function are documented and validated, then the next step would be to develop a personnel security flowchart or charts.

In this case, separate flowcharts are developed due to major differences between the two functions of investigations and WPV prevention. They are connected in some ways; for example, if an employee threatens to or actually performs acts of violence, one of the after-action tasks must be to determine how the individual made it through the prescreening process.

SAMPLE PRE-EMPLOYMENT AND BACKGROUND INVESTIGATIONS METRICS CHARTS

Based on the flowcharts, the CSO targeted several areas for security metrics management. This was done *not* to determine if the process was operating as effectively and efficiently as possible. That would be premature. Based on targeted areas identified by analyzing the process as shown in the flowcharts, the CSO wanted first to determine the quality of the process (how well it was being performed) and the number of background investigations being performed. This would size the statement of work. Furthermore, the CSO needed to understand what all of this was costing the company.

Before trying to make changes to a process with the goal of making it more effective and/or efficient, it is first necessary to have something to compare it with. The purpose of any measurement is to be able to use that metric as a point of comparison. That comparison may be against past performance (trend), or a comparison of output against money spent (efficiency). For example: doing the same amount of work from one year compared to another but at a lower cost. This suggests an improvement in efficiency.

The first thing the CSO noticed on the flowchart was some prescreening forms that were filled out by the applicant and checked by the HR employment office staff were sometimes returned by the security staff to HR due to errors, such as incomplete information, time gaps in applicant background (education and employment), or a failure to obtain the privacy waiver signature of the applicant, etc. This failure causes a delay in the start of the process. Implementing a quality check early in the process should prevent these errors and eliminate the interruption to the start of the process.

This simple tracking tool allows the security professional or process owner to immediately see a problem with the process. Continuing with this part of the process, what would you, the CSO, like to see as a follow-up chart?

Remember, your goal is to make this entire process as effective and efficient as possible. With that goal in mind, wouldn't it be nice to know how long it takes for the pre-employment form to be corrected and returned to personnel security for processing? Of course it would, and why is that important? It is important because the potential employee has been identified and the longer it takes to process the pre-employment investigation the longer it may take to get that applicant on-board with IWC as an employee. This delay translates to work not being done and the organization being less productive.

It may very well be that others are required to take up the workload due to a shortage of staff, which in the long run adversely affects employee morale. Poor employee morale is one of many conditions adversely affecting employee productivity. Add the problem of increased workload leading to additional errors, and the problem becomes worse.

What does all this mean? The cascading effect of the lack of quality control in properly completing the forms—"doing it right the first time"—can be very costly.

Once you identified this problem, you may want to identify the costs associated with it. For example, determine the starting salary of the new employee and multiply that by the number of days the form is delayed in the process. That will provide an estimated cost in productivity loss for that new employee, assuming the person was hired based on an investigation indicating no employment-related issues.

You may think that is stretching this thought process a little, but it really isn't. That is because IWC needs this employee to work as soon as possible and is willing to pay a certain wage for the productivity of that employee. So, it makes sense to address the issue from this point of view.

One can also take this analysis further and interview personnel in the organization to which the new employee would be assigned and determine how they are meeting their goals without this or several new employees. However, that may be far too time-consuming (and only provide a greater level understanding of an already known problem) and at this point in the security metrics management application to this issue, it is unnecessary since the primary goal is not the impact on the other employees but eventually providing a more efficient and effective process.

So, as the CSO monitoring these metrics, what would you do to correct the problem, or at least get the rejection numbers lowered?

You can go to your boss, the vice president, HR, and explain the issue. However, is that really necessary at this point? Were you not hired to solve problems like this yourself? Part of your role as the CSO it to identify problems and find solutions. The best course of action is to meet with the HR manager responsible for that part of the process and explain the problem. When doing so, do not come across as complaining but as one who says that "we" have a problem. Explain the problem using your security metrics charts and then propose a solution. Why should you propose a solution?

Because you own this problem, as this process is in place as part of the CAPP and you are responsible for leading the CAPP and that means all its policies, procedures, plans, and processes. The HR manager is just following IWC and CAPP policies by being part of the pre-employment investigative process.

The problem, according to the HR manager, may be that there are several new employees involved in this process and they are learning on the job. However, it may very well be that they are understaffed so the three people handling all these applications instead of the usual five are overworked and producing lower quality results. As the CSO, you may not want to ask the HR manager why more HR specialists are not hired. The reply may be that they are awaiting the completion of the pre-employment investigation! What would be your proposed solution?

The immediate solution may be to hold a training class for all personnel involved in the process. The pre-employment investigation form could be reviewed, explained, and discussed step-by-step or even revised. The HR and security personnel involved would also get to know each other better and identify where the people working the process see issues. Parts of the form may be vague and they may agree that some changes to the form would be helpful. Also, they may be able to change their daily approach to problem solving by engaging others working on the process in a productive way (a phone call from the personnel security specialist would help speed up the process instead of just mailing the form back to HR with a form letter on top saying why it is being returned).

Once any form changes were completed and the training session completed, would you drop the security metrics management of this process? As the CSO, you would not want to eliminate this metrics because it is important to see if the problems identified and allegedly corrected are in fact the systemic problems that caused the process to be ineffective in the first place. Moreover, the process needs to be monitored to determine if the changes made were effective and what is the cost of the process.

Keep in mind, the purpose of measuring any process is to have a means of evaluating the performance of that process. Knowing about a problem or change early will lead to early corrective action, thereby reducing the amount of lost time and money.

One important aspect of this issue has not been discussed and that is, how do you collect the data to begin with? The data should be collected where the problem was identified. In this case, it would be collected by the personnel security staff that check the form and begin the pre-employment investigative process.

So, once the "data collectors" are identified, how is the data to be collected? The easiest way is to use a manual log or computerized database or spreadsheet. Using a database or spreadsheet to input the data may work the best, as that may make it easier to compile the data for analyses and also for making out metrics charts to show that compiled data in a meaningful format for analyses and briefings.

As the CSO, what data would you like to have collected? Well, in order to provide the data that will serve the most good for targeting the problem, the data collected should at a minimum be:

- Time delays in processing the forms
- Persons involved in the process (to identify those making the errors as candidates for additional training)
- Name of the applicant
- Future organization of the employment applicant to determine which departments are most affected. Possibly their needs might be prioritized as far as which investigations get done first. As an example, maybe a spreadsheet that can be used to track the receipt and status of all pre-employment investigations. What other types of charts can you think of that may work to help you display and analyze this information.

Using this spreadsheet, data such as average dates of processing time within the personnel security department, number of requests processed, how many from each department, whether or not derogatory information was found, and so forth, can all be compiled, analyzed, charted, and reported from this one form. It can also be used to identify the costs of this entire process, as well as portions of the process.

This metric allows for the development of an average expectation of the percent of investigations that produce derogatory information about a potential candidate. This metric will be useful to view as a trend. If it changes significantly, it may indicate the applicant pool needs to be reviewed or changed. It may also lead to further analysis. As the CSO, what other relevant metrics would you like to have available?

Once the data is collected on the time it takes to conduct a pre-employment investigation (from the time the correct pre-employment investigation form is received by the personnel security office staff to the time the investigation is completed and the results forwarded to the HR staff—the total time spent conducting the investigation) and reporting the finding cannot only be tracked but the time spent by the personnel security staff can be converted to costs. In addition, the data collection should be in such detail as to track all the tasks associated with that process:

- Time it takes from receipt of the form to the time it is assigned to an investigator.
- The time it takes the investigator to conduct the investigation.

- From the time the investigation is completed until the time the report is produced.
- From the time the report is produced until the time it leaves the personnel security office.
- From the time it leaves the personnel security office to the time it reaches HR.

Each detailed step in the process can also be equated to time spent and all time spent can be converted into costs.

As the IWC CSO, can you identify other security metrics management data showing costs that you would like to have charted for analyses and executive management briefings?

Here are examples of other costs that could be identified for further analyses and use in analyzing the process and identifying more efficient and effective pre-employment investigative methods and measures:

- Average cost for each investigation.
- Individual total costs for completing each standard investigative step, e.g., credit bureau check, reference interviews, public database check, etc.
- Average costs for completing each standard investigative step, e.g., credit bureau check, reference interviews, public database check, etc.
- Average cost (time and USD) of the personnel security staff effort in the total process of an investigation.
- Total costs (time and USD) of the personnel security staff effort in the total process of an investigation.
- Average costs (time and USD) for each detailed step in the pre-employment investigative process.
- Total costs (time and USD) for each detailed step in the pre-employment investigative process.

Each of the steps in the process could be identified in a detailed flowchart and subject to further analysis to include measurement. The reason is that each step in the process could be made more efficient. For example: The investigator conducts the checks required, makes notes accordingly, and when investigative work is complete, prepares a final report of the results of the investigation. Suppose that the investigator used a digital recorder for use on the road when conducting the investigation and also had a voice recognition software program on the office computer.

The investigator could dictate the results into the recorder and, through the USB port, have it downloaded to the office computer and automatically typed. In addition, using the computer microphone, the investigator could also dictate the investigative results as the investigation was progressing. This would also be automatically typed—without mistakes and with the potential for greater speed and accuracy. This is an example of using technology—information technology—to produce a better and more efficient process.

This method may be cost-beneficial by shortening the time it took to conduct and report the investigative results back to HR. However, this is a possible solution that should only be considered as part of a project plan to identify and analyze the time spent conducting investigations. The reason that this should not be prematurely considered is that you may not have examined all aspects of this process in sufficient detail and prematurely identified the problem; however, it may have been more of a symptom and not the systemic problem. You would have implemented a solution costing a considerable sum of money to purchase the software and hardware and train the staff, only to find out that the time it took to conduct the investigations and report the findings were the same or on average may have taken longer.

It is not uncommon for project teams to look for quick solutions, using technology, but end up with a costly change producing little or no improvement.

> Understand the process and its weaknesses before making changes.

We want you to avoid that unwanted condition of "getting your daily exercise by jumping to conclusions." This may be such a case, so caution is warranted.

WORKPLACE VIOLENCE

Workplace violence can be perpetrated by employees, visitors, customers, or those who choose to engage in criminal behavior. A situation where two employees engage in a fistfight in the IWC parking lot is also an act of workplace violence. Workplace violence is not limited to acts of physical violence alone. Threats of violence or the fear that one may be subjected to violence is a form of workplace violence. A threatening environment can be just as disruptive and damaging as any acts of physical violence. Conditions that create an environment of fear can be as damaging to the workforce and individuals as actual acts of physical violence.

As the CSO at IWC, you are interested in using security metrics management techniques to determine not only the costs of the IWC workplace violence program (at least the costs to the security department), but also the costs of violence in the IWC workplace in terms of lawsuits, lost productivity due to disruptions in the workforce, the time it takes for additional training following an incident, and the cost of an employee being terminated and cost of hiring a new employee. Workplace violence incidents have an adverse impact on any company in many different ways.

In the effort to better understand the effectiveness of such a program, what types of data would you collect and what measures would you use to help you better understand the process and be able to do analysis work?

- What data is needed?
- What do you want to know?
- How often would the data be collected?
- Who would collect the data?
- When would they collect the data?
- Where in the process should the data be collected?
- How is the data to be compiled and reported?

> Since workplace violence is often a product of a hostile, unsafe, or unfair work environment, identifying the various indicators of a hostile, unsafe, or unfair workplace is the best place to start.

Assessing the condition of the workplace can be accomplished by conducting workplace climate surveys. The CSO must work with HR to develop and administer such a survey. How employees respond to questions related to fairness (usually in terms of how they are treated by management and other employees) workplace safety, and working conditions, will produce enough information to make a determination of the general conditions of the workplace. Should there be negative indicators developed during the survey, senior management must be engaged and a plan developed to address the identified issues and change the negative workplace climate.

The most obvious measure for assessing the effectiveness of WPV prevention is by measuring the number of incidents. This analysis can be taken further by measuring the number of averted incidents through use of a referral to Employee Assistance Program (EAP) process. When management and employees are sufficiently trained to recognize indicators in behavior as having the potential for violence and they intervene during the early stages of behavior change, it can be assumed that a potential WPV situation was averted. As complicated as this may sound, the CSO working with HR and EAP professionals should be able to develop a training program for managers and employees and a means of measuring the effectiveness of a referral process in terms of WPV prevention.

CASE STUDY

The IWC CSO was concerned with the security department's share of the costs of the WPV prevention efforts when the CSO found out that this was a recently started program brought on by an assault incident when an employee subjected to a layoff, as a result of a reduction in force, acted out violently. The employee later sued not only for wrongful discharge but also because IWC had not prepared him for such an event and if it had, he never would have assaulted his boss. The employee believed that IWC did nothing to assist the employee in the transition from IWC to another job opportunity. In essence, the employee claimed there was no support structure in place to assist employees affected by work force reductions.

As the IWC CSO, do you think the workplace environment was as good as it could be during this difficult period? Could IWC have taken actions to mitigate the pain and suffering experienced by employees being terminated? Perhaps the affected employee felt mistreated, with dignity damaged and resentment of the perceived callous attitude on the part of the company so, under frustrating circumstances, lashed out.

After this incident, IWC's executive management immediately wanted a total program review to assess the workplace environment and determine what needed to be done to improve IWC's ability to prevent incidents of workplace violence. Naturally, this effort will cost time and money.

Although the CSO understood the need and the urgency, budget was a factor. Therefore, the CSO implemented a project to identify and track the security department's costs so that during the next budget cycle, the CSO could use the security metrics management program (SMMP) and its associated metrics charts to justify the new security budget. Of course, the metrics charts relative to this effort would be included.

Beginning with a high-level flowchart, the CSO's analysis concluded that for several key targeted areas, metrics compilation and charting would be completed in time for the new budget talks in approximately 6 months.

However, not waiting for the budget talks, the CSO took a look at the requirements and then talked to the manager responsible for the new-hire orientation. They reached agreement that the security briefer would train several of the HR personnel who were responsible for providing other briefings during the new-hire orientation day, as well as provide them pamphlets on the anti-workplace violence program and the briefing slides that were used by the security briefer. HR would then do the security orientation briefing, including the WPV prevention.

The security department thus saves the 1 h briefing time per week, as well as the preparation time and time needed to get to the briefing room, a total of 2 h/week. This savings was calculated as: 2 h/week × 50 weeks/year = 100 h a year of time saved and at $20/h including benefits = 2000 USD. Yes, this may be considered a transfer of costs and not a savings per se, but it does help the CSO and still gets the job done.

This metrics chart shows what type of workplace violence incidents may be occurring. This will help direct the intervention and training program for WPV prevention to emphasize those areas which may help minimize the probability for such incidents. Furthermore, IWC managers could be further briefed on what indicators to be alert for in any employee that may be indicative of such contact so that it could be prevented through intervention by management with the employee before escalation. Prevention is the key.

Follow-up metrics would include one showing the time spent per investigation. Another would show the action taken by HR based on the investigative findings, as well as costs throughout the process.

Is there a trend developing? Are acts of hostility increasing? Is there a behavior change taking place in the workforce? All are questions that must be asked when the number of incidents or threat of incidents increases.

Regardless of the type of incident, an SMMP will assist in determining changes and trends. In addition—and somewhat of an unfortunate reality—the security investigators can prepare for the additional workload.

SUMMARY

Under administrative security is the function of personnel security. Personnel security has two primary functions: (1) pre-employment and background investigations and (2) WPV prevention program support.

Using security metrics management, the effectiveness of both programs can be assessed. Moreover, the cost of preventive measures can be measured, as well as the consequences of having and not having effective preventive measures in place.

CHAPTER 4

Information Security

CONTENTS

INTRODUCTION

Since information is one of today's most valuable assets for any organization, business, or government and is one of the "triad" of valuable assets—(1) people, (2) information, and (3) physical property such as facilities and equipment—it must be protected.

The protection of information in this age of electronic and digital information is more important, and more complex, than ever before. The loss or theft of information critical to a corporation's products, methods, or processes may be devastating. In this age of global competition, the importance of implementing a comprehensive program for information protection is critical and cannot be overstated.

THREE BASIC CATEGORIES OF INFORMATION

International Widget Corporation (IWC) divided its information into three basic categories:

1. Personal, private information
2. National security (both classified and unclassified) information[1]
3. Business information

Personal, private information is an individual matter, but also a matter for the government and businesses. A person may want to keep private such information about themselves as their age, weight, address, cellular phone number, salary, and their likes and dislikes. Many nation-states have laws that protect information under some type of "privacy act." In businesses and government agencies, it is a matter of policy to safeguard certain

[1]National security information is discussed in Chapter 16, Government Security and will not be repeated here.

Security Metrics Management. DOI: http://dx.doi.org/10.1016/B978-0-12-804453-7.00004-5

information about an employee such as their age, address, salary, etc. The IWC Chief Security Officer (CSO) understands that, although the information is personal to the individual, others may require that information. At the same time, IWC has an obligation to protect that information because it is considered to have value. Therefore, this information is categorized as a vital asset requiring protection under the corporate assets protection program (CAPP).

Business information also requires protection based on its value. This information has been categorized at IWC as:

- IWC Internal Use Only
- IWC Private
- IWC Sensitive
- IWC Proprietary
- IWC Trade Secret

This information must be protected because it has value to IWC. The degree of protection required is also dependent on the value of the information during a specific period of time.

SECURITY DRIVERS

What are the security drivers that create the need to protect information assets? Information is a valuable asset as was stated earlier. However, not all information is considered an asset. Since information, as it is categorized at IWC as stated above, defines what is important, we can deduce that the information that requires protection is that information that is needed by IWC and which would cause harm to IWC if not properly protected, and became available to unauthorized people without the need to know (NTK).

Therefore, information assets that must be protected are those information assets which, if not adequately protected, would be in violation of laws, regulations, management direction, corporate policy, ethics, and otherwise would adversely impact IWC. These are the information assets protection drivers.

INFORMATION SECURITY PROCESS FLOWCHARTS

Information that requires protection must come from somewhere. There are two primary sources of information that requires protection:

1. Information that is born from the brains of IWC employees who develop such information in the course of performing their IWC

duties. That information then becomes the exclusive property of IWC and is to be protected and used as directed by IWC's CAPP.

2. Information provided to IWC by customers, associates, contractors, and others with a stipulation that the information provided must be protected from unauthorized access, damage, modification, destruction, or theft. It is to be used only as they direct its use. This information often is provided in writing in the form of contractual requirements.

The CSO's information security project team developed a high-level flowchart that is to be used to provide a "picture" of information flow. Based on that flowchart, metrics targets can then be identified.

When looking at this flowchart, some concepts relative to security metrics management should become apparent. The process to categorize the information is time-consuming, as are the protection processes that follow. One has to be sure that information is adequately protected, which means employees understand the sensitivity of information and they take actions such as locking documents in proper containers, maintaining an accountability record for the most important information such as proprietary research information, not discussing sensitive information with people who have no NTK, and many other measures.

And as previously noted, a process must be in place to provide a time limit for protecting information, to include a periodic review to determine if protection is still required, or if less protection can be implemented, which will save some resources—for example, safes or accountability controls. All information is perishable or has a "shelf life." It's important not to protect information for periods that are far too long or far too short.

So, in this case, the flowchart allows the CSO to analyze the information protection process from "cradle to grave." Of course, your flowcharts would be in much greater detail so that you can do an in-depth analysis.

One other thing to keep in mind is to try not to do too much at one time. Some tasks can be done by the project team in serial order and other tasks in parallel. For example, after you analyze the process supported by the flowchart, the next step would be to interview not only those that perform the information security duties but also the IWC employees who are affected by the information assets protection policies and procedures. You want to know if the IWC Information Classification Guide is useful. If not, why not? You may need a separate project to address that issue.

As the CSO, you want to know how much equipment is used throughout IWC to protect information. You want to break that down by location and type of equipment, such as safes, key-lock filing cabinets, or information system access control software, as well as costs.

This process is the beginning of the security metrics management process for the information assets protection function:

- Identify the information assets protection drivers.
- Develop and analyze the information assets protection processes flowchart.
- Target areas for data collection with the ultimate goal to make the process more effective and efficient.
- Develop a method for data collection.
- Collect and collate that data.
- Analyze the data.
- Develop security metrics management charts as a management tool and also to brief executive management on the matter or as part of an overall CSO briefing to executive management.
- Develop and implement projects to increase effectiveness and efficiencies of the processes and subprocesses.

WHAT, WHEN, WHO, WHERE, HOW OF DATA COLLECTION

As part of this project, one must determine:

- What data will be collected?
- At what points in the process will the data be collected?
- Who will collect the data?
- When will they collect the data?
- How will they collect the data (manually or through an automated process)?

Once that is decided, a pilot subproject or major milestone task should be initiated under this project—to test the collection process. After all, you would not want to establish the collection process and ignore it until the first data collection is completed after 1 month or more, and then find out that the data collected is not what you had in mind. So, not only is the data collection process to be tested but those involved in the collection process must understand what they are doing and why they are doing it. Furthermore, they *must* be directed to notify the data collection project lead, their manager, or you if they have any questions or the process is not working based on its objective.

As far as data collection from employees, one can send them a survey form to complete. Be sure it is clear why you want the data and provide them an easy process to report the survey—e.g., place it online, send it out by e-mail, and have them reply by e-mail.

Also, you must determine if the survey information is to be tabulated manually and input in a database, spreadsheet or if a program is written and placed online that will allow for the automated collection of the data, ready for analysis. Remember, all this takes time and time is money.

Analysis is needed but caution: beware of analysis to the point of paralysis.

You would undoubtedly want to present your charts in color and probably view them online at your desk. However, remember that they may be reproduced as black-and-white charts. So, always be sure to check them not only for accuracy and to understand what they are showing, but maybe just as important, are they clear to the viewer?

As the CSO, you will probably be responsible for providing the handouts and, as cheap as color printing has gotten, colored handouts would not only look more professional but also allow easier viewing of the data.

Compare the views presented in the previous charts. Which is better? Why? Would you choose another style? If so, ask yourself why? Always ask yourself why you chose the chart style you have over all the others.

A CASE STUDY

If you recall, the CSO found that IWC had five different types of business information. They were:

1. IWC Internal Use Only
2. IWC Private
3. IWC Sensitive
4. IWC Proprietary
5. IWC Trade Secret

The CSO also knew that this information must be protected because it has value to IWC. The degree of protection required is also dependent on the value of the information during a specific period of time. So, although each category requires some degree of protection, that is dependent on the information assets protection policy. No matter the category, the information assets that are being protected are all to be protected from those who do not have a NTK for that type of information.

So, the logic is that the degree of protection mitigates to various degrees the risks to the information asset being protected. But still, the goal is not to give unauthorized personnel access to it, regardless of its classification.

If you were the CSO, what would you do to lower the costs of information assets protection? Yes, there are many possibilities; however, there is one basic possibility that may be the easiest to deal with and that is eliminating some of the information classifications. Review the categories of information and that information that is not sensitive, or may be only somewhat sensitive for a very short period of time, could be protected as routine information requiring no additional controls. In essence, the universe of information requiring protection would be shrunk to include only the most sensitive information. Yes, there may be a slightly increased risk of exposure to some information, but that may be offset by the economic savings of reducing the amount of information requiring additional (and continuous) protection. If you were the CSO and wanted to look at that possibility, how would you go about doing that and in what order?

Here is one possible scenario:

- Initiate a project plan for tasks that have an objective, beginning and ending date and are classified as projects.
- Identify project team members.
- Brief them on the objective of the project.
- The goal is to find enough commonality among the categories to reduce the five to two—one for sensitive, trade secret, and proprietary information and one for all other IWC information that is not to be released to unauthorized personnel.
- Get their input.
- Identify tasks and assign them.
- Identify the policies, procedures, and processes used for each category of information assets being protected.
- Identify commonalities of protection among the five categories.
- Identify the costs of that protection (for each category of information).
- Identify the different types of physical containers approved and used to protect each category of information.
- Determine how many containers there are within each IWC department and compile by region and then total.
- Determine if there are any security drivers external to IWC that require the classification (naming convention) of certain information.
- Determine if there are any security drivers external to IWC that mandate specific protection mechanisms.
- Based on the data collected as to the costs for protecting certain information in a certain way determine if you could reduce those

categories to two and recalculate the protective measures costs, assuming that the minimum protection requirements will be used.
- Brief management using the security metrics management charts.
- Gain their approval.
- Establish an implementation plan.
- Implement the plan.
- Monitor the process and costs thereafter using the security metrics management tools.

SUMMARY

Information assets are one of the three types of assets requiring protection. Like all other types of security functions, information assets protection is costly and some of the costs are unknown.

Even information that requires protection today may not require protection tomorrow, as information is time-sensitive. If you protect information no longer requiring protection, you are wasting valuable corporate resources. If you misclassify information, you may not be protecting it as you should or over protecting it. Either way, it is costly.

Using security metrics management techniques:

- identify security drivers,
- identify information assets protection processes,
- flowchart the processes,
- identify target areas of opportunity to reduce costs in terms of money and time,
- establish projects to quantify costs, analyze the processes, and develop modifications to the processes that will make them more effective and efficient.

CHAPTER 5

Security Compliance Audits

INTRODUCTION

CONTENTS

The security compliance audit (SCA) function is an integral part of the administrative security's organization. Some may see this as a conflict with the corporate audit function; however, while it may give that appearance to some, it actually complements the audit department's function, as regards assets protection compliance. They are two separate and distinct organizations, each with a different statement of work but using similar processes and sharing overall objectives.

SCA provides a look at compliance from the perspective of those who have an expertise in security regulations, policy, procedures, processes, and practices. It is always useful to have a second "set of eyes" looking at how a security operation works.

> The "corporate audit" function traditionally focuses on compliance while the SCA's main focus is on ensuring that assets are properly protected, regardless of compliance with the corporate assets protection program (CAPP).

Internal corporate audit teams generally do not have security professionals as permanent members of their team. Thus, the SCA serves a worthwhile function in the goal of protection of corporate assets.

While the corporate audit department is responsible for determining compliance with all of the corporation's policies, procedures and contractual agreements, and external regulations and laws, the emphasis and focus of the SCA, as alluded to earlier, is limited in scope to the protection of the corporate assets.

Security Metrics Management. DOI: http://dx.doi.org/10.1016/B978-0-12-804453-7.00005-7

The SCA can be described as the security department's "internal audit" process. The SCA should focus on the following:

- Assessing whether or not assets protection regulations, policies, and procedures are actually useful in supporting the protection of assets and, if so, are they not only effective but efficient.
- Testing assets protection systems and processes to ensure they work as expected.

Furthermore, the SCA process may be used to assess compliance of associated suppliers, subcontractors, and partners that have access to sensitive proprietary assets of the corporation—in this case, those assets of the International Widget Corporation (IWC).

Like all other security functions, in order to begin a security metrics management program (SMMP) for this function, a list of drivers and the associated flowcharts must be developed.

SCA DRIVERS AND FLOWCHARTS

The primary driver for this security function is the IWC CAPP itself. As we stated in earlier chapters, the CAPP is driven by the other security drivers, such as laws, regulations, IWC policies, procedures, contractual agreements, etc.

The CAPP drives the SCA function because the CAPP must be complied with by all IWC employees and associates according to their contracts. How does one know that each department and each employee in each department is complying with the CAPP and therefore fulfilling their obligation to protect IWC assets? The answer is that no one really knows unless someone checks to find out if everyone is complying with the CAPP; thus the need for the SCA function.

In addition, the SCA also tries to determine if the protections in place are actually protecting assets. For example, if the CAPP is being complied with, that does not necessarily mean that the assets are protected as they should be. There may be a flaw in the asset protection defenses. A proactive view—for example, testing attack methods against assets—is an SCA function to better determine the protection of assets regardless of policies, procedures, and such in place.

SCA METRICS

The SCA function, to be effectively and efficiently managed, requires that there be metrics in place to track performance. Measures such as the number and scope of security audits, project costs, resources used, and to the extent possible, benefits derived. Benefits derived are the positive attributes of the effort. For example, if during the SCA process it is determined that a portion of the CAPP is out of compliance with recent regulations, or operating ineffectively, there is value in discovering this condition and correcting it before there is a system failure or before an external agency discovers the problem or problems. Perhaps penalties can be averted by discovering and correcting problems before they reach a more severe level.

One of the SCA goals is to develop and maintain an annual schedule of all required security audits, tests, and evaluations. This schedule should be published and distributed to affected parties. Moreover, it should be developed with input from all affected organizations. The goal is by the end of the year to successfully complete 100% of all scheduled SCAs. Another and maybe most important goal is to have IWC and all departments operating at a satisfactory level of compliance. This effort should prepare IWC for all external audits and help achieve a 100% satisfactory compliance condition. Operating out of compliance is a high-risk condition and the SCA process helps the Chief Security Officer (CSO), and thus IWC, avoid that condition.

The primary effectiveness measure of the SCA's program occurs when all external auditing or inspecting agencies, after conducting an audit or inspection of IWC, and possibly the security department, at any or all locations, conclude the operations to be compliant. The achieved condition of compliance, along with measurable reductions in loss (or damage) of IWC assets are the real metrics on the effectiveness of the SCA.

Of course, in terms of the SCA process itself and ensuring its effectiveness and efficiency, it is necessary to measure it. All SCA processes and products (security auditing processes, checklists, tests and evaluations, and reports) should be analyzed for efficiencies.

Using process flowcharts to understand how each works, tracking costs and use of resources will enable the CSO to better understand what it costs to manage and maintain an effective SCA. Moreover, measuring SCA performance allows the CSO to ask the questions, the answers to which will lead to a more effective performance. For example:

- Is there really a benefit to IWC in providing SCA tests and evaluations? In other words, are fewer assets vulnerable to theft or unauthorized destruction due to the process?
- Does the SCA produce value greater than it costs to maintain?

- Does the use of the SCA processes and products lead employees to better protect the IWC assets?
- If not, what other purposes do they serve?
- Are they (SCAs) a "nice-to-have" (producing little or negligible results) item with no visible return on investments?

An SMMP will help answer the above but also assist the CSO in making better management decisions. In times of corporate frugality, such a function may seem somewhat redundant when one considers the CAPP is subject to other audits—internal (corporate audit) and external (outside agencies such as those with regulatory authority over business and industry and those specifically engaged in a contractual relationship with IWC). However, if measurable results demonstrate the success of the SCA as a value-added security function, it will be harder to make a case that the SCA function should be eliminated.

Without measurable results, a great deal of pressure may be placed on the CSO to eliminate the SCA function because of using finite resources for an undefined or ill-defined result. That may also be used as the rationale by IWC executive management or others for eliminating the security function when the real reason is that the SCA reports have made them look bad due to their noncompliance with some portions of the CAPP as required.

Regardless, if the SCA is being done, it must be done as cost-effectively as possible. There may be other methods of implementing an SCA and therefore the following questions should be asked:

- Can such a function be done more efficiently by having each IWC department conduct a self-audit once a year?
- If so, what are the cost–benefits?
- Can they be done as effectively or is there a risk of the "fox watching the hen house"?

One key factor to always remember: When approaching such matters, the CSO should look at them with a holistic, or IWC-wide, approach. In this case, are costs in terms of resources used or productivity losses actually being reduced, or is there just a transfer of costs from one function or department to another?

Whether a security audit is being accomplished by another internal department employee or an IWC security specialist, if the hours spent are roughly the same, then from an IWC viewpoint, the costs may be transferred from the security department to the other IWC department. However, although the costs in time and effort are the same, due to differing levels of expertise, there may be hidden costs buried in the quality of the effort. Someone with specific security expertise is more likely to understand the nuances of the security function and may be better able to identify problems than someone with little or no security (assets protection) experience.

THE WHO, HOW, WHERE, WHEN, WHY, AND WHAT OF SCA METRICS TRACKING

As with all other security functions, where security metrics management techniques are applied, the CSO must determine who performs what data collection and tracking task. What is tracked, and who tracks it, along with the when, where, how, why, etc.

Who should track the functional metrics input data? As with all security functions, the tracking of the data and data collection should be done by the person at the lowest level who is responsible for the day-to-day SCA activity.

What to track? To begin with, all major tasks should be tracked and gradually, data collected in more detail as one moves through the work-breakdown structure of the SCA. To start with, the tracking and data collection should focus on basic data. That includes the number of security audits conducted, the security audit results, and their impact on assets protection and of course their associated costs.

Why track it? Because the SCA is an important security function, expending much employee time and resources, so it must be efficiently managed. All subprocesses (e.g., employee time worked, travel time and other expenses, report writing, production and distribution of reports) must be identified and measured. Once these factors can be quantified and qualified, then the process of analysis can begin to determine more cost-effective and efficient ways of providing the SCA service and support to the CAPP and thus IWC.

How to track it? To begin with, the tracking of data collection costs and lost productivity of workers can be accomplished by tracking the time spent by security specialists in doing the security audit preparation work and actually conducting the security audit along with the time spent by employees within a targeted department preparing for and interacting with security audit team members during the security audit. How much time each involved employee spends, the resources they use, and the amount of time not spent doing their normally assigned tasks are all part of the cost of the SCA.

Once the SCA is completed, then the time involved in each step of the process (using a detailed flowchart of that process as a guide) can be entered into a spreadsheet and the total time spent conducting the SCA can be calculated, as well as the average time spent by each employee. This can later be compared with the SCA results over time, as well as whether or not assets subsequently are found to be vulnerable or lost.

Such losses, for example, can then be used not only as part of a security education and awareness program briefing, but also as part of an SCA evaluation. In other words:

- What happened after the SCA was conducted in the department that contributed to or failed to prevent the loss of an important asset?
- Could it be the SCA was not done as effectively as possible?
- Was something missed?
- If so, that would require correction and changes in SCA procedures and related processes.

> The process of measurement leads to continuously looking at data in an attempt to discover areas for process improvement.

When to track it? Various aspects of the SCA should be tracked by updating all related spreadsheets and databases as each SCA is completed. Maintaining accurate and current records is critical. Trying to reconstruct data at a later date usually ends up with an inaccurate depiction.

Where to track it in the functional process? Each SCA's data would be collected as each stage of each of the SCAs progressed. The collection points should at a minimum be at all of the primary transaction or action points noted in the process flowcharts for the SCAs primary and subprocesses.

Once data collection is complete, the data must be depicted in a useful form. In analyzing the data depicted in graphic format, the CSO can begin to understand just how the process is really working. Trends and anomalies in performance, costs, and use of resources can be examined. Positive and negative assessments can then be made, followed by decisions to implement process changes as appropriate. Costs in terms of hours expended and materiel costs can be shown on spreadsheets and charts for analysis.

Changes to the SCA process can be implemented on a trial basis based on the suggestions by the department personnel and security specialists to determine if the changes would result in a more cost-effective SCA program. However, changes made based on data analysis almost always produce better results. This is particularly true when attempting to improve productivity—real data is necessary as a point of comparison. Anecdotal data (we think we are doing a better job) is interesting, sometimes informative, but not useful for real analysis.

SCA CASE STUDY

The CSO was asked by the boss, "What is this security compliance audit function and what is it used for?" As a CSO, how would you answer? The IWC CSO thought about it and told the boss:

The SCA function complements all other assets protection functions. It provides a second look at how the CAPP, its assets protection policies, procedures, projects, and processes are working. Moreover, it is a mechanism that can be used to determine how the entire company is fulfilling its responsibility to protect the corporate assets.

The SCA process is best conducted by the security organization, but it could be accomplished at the department level. Let's assume that a self-audit (that is each individual department) program was implemented and the SCA function discontinued. The work would still be accomplished but would it be of the quality needed to ensure the CAPP is effective and efficient?

Are non-security "auditors" as effective as security professionals in conducting security audits of the IWC assets protection processes? They do provide that "outside" perspective, but their lack of security and of assets protection knowledge may lead to flawed or less than optimal audits. In addition, if one is doing a security audit of their own department, can they be objective to the point of issuing a marginal or unsatisfactory rating if that was the result of the security audit?

Each approach brings its own advantages and disadvantages. Only after careful analysis can the CSO make a determination as to which approach is the most effective and efficient. However, other mitigating factors may drive the decision one way or the other. If there are resource issues challenging IWC, there may not be sufficient security professionals to do this work. It may be necessary for the functional departments to conduct their own assessments, or not do them at all. Further, the amount of resources dedicated to the SCA process may be a factor in how much risk IWC executive management is willing to take. If external oversight agencies are not that demanding, greater risk acceptance may be an option ("Hey, they aren't closely watching us so we can let things slide a little."). However, that is not viewing the matter from an assets protection viewpoint, is it?

In any event, the CSO should drive decisions regarding the SCA based on data collected and analyzed to better assess performance. This is the only logical departure point.

The IWC CSO then went on to show the boss the SMMP data collected and their supporting charts to further explain the SCA function.

SUMMARY

The SCA function is an integral part of the administrative security organization. It is more than security auditing for CAPP compliance as it is driven by the goal of ensuring all defenses are in place and compliance is achieved in accordance with the security drivers—that is, external regulations and laws, as well as internal policies, procedures, projects, plans, and processes. The ultimate goal is the efficient and effective protection of corporate assets—essentially, achieving an effective CAPP.

The SCA process can be a very costly function in terms of resources used by the SCA team and time spent by IWC department employees who would take time out from their normal jobs to support the security audit.

Identifying the costs and benefits to IWC requires measurement. Measurement must be conducted in an organized and comprehensive way. Data has to be collected, depicted in a meaningful and understandable form, and analyzed. The analysis provides the CSO with sufficient information to cause process changes as needed. Changes that are made must drive improvements in terms of effectiveness and efficiency.

Collecting data and depicting it graphically is not a difficult task. The basic charts to use for simple graphic depiction include the organizational chart, process flowcharts, and security metrics management charts such as those showing the number of security audits conducted, their results, and the correlation of those results with the noncompliance incidents and investigations related to the loss of IWC assets, as well as incorporating the time being spent by all involved in the SCAs.

CHAPTER 6

Security Education and Awareness

CONTENTS

INTRODUCTION

The security education and awareness program (SEAP) mission is to ensure all International Widget Corporation (IWC) employees and support personnel (e.g., in-house contractors, suppliers, and partners) are aware of their responsibilities to protect IWC assets. Another objective of the SEAP is to teach them how to protect those assets. Through periodic security briefings, reinforced with the development and distribution of security awareness materials, the Chief Security Officer (CSO) must continue to work to raise the level of security consciousness within IWC.

In addition, the IWC CSO has charged the security staff supporting the SEAP to manage this function as efficiently and effectively as possible. This includes ensuring the security staff is well prepared to develop and deliver security training materials specific to the needs of IWC. However, for our purposes in this chapter, we will concentrate on the two functions, security awareness and how to protect assets.

Once all the administrative security tasks, such as policies, procedures, plans, and processes, are in place, those that are expected to comply with them must know about them. After all, they would be useless if IWC employees and other personnel did not know that these policies, procedures, plans, and processes even existed or what the employees and others need to do to comply with them.

The entire IWC corporate assets protection program (CAPP) consists of security or assets protection "layers." One of the "foundation" layers is employee vigilance and understanding as to how to protect corporate assets. That understanding and hopefully the motivation to protect corporate assets are learned and developed through the use of SEAP tools (briefings, videos, pamphlets, and other security training and awareness material).

Security Metrics Management. DOI: http://dx.doi.org/10.1016/B978-0-12-804453-7.00006-9

The SEAP is used to make the IWC employees aware of explicit assets protection policies, procedures, and how to comply with them. The objective, of course, is for everyone at IWC to protect corporate assets, supported by the employees having a clear awareness as to why and how to properly protect the IWC assets.

The IWC SEAP is based on effective communications with constant feedback, supported by measurement (security metrics management program (SMMP)), to determine if the security education and awareness training program (SEATP) is meeting its established goal of lowering assets' threats, vulnerabilities, risks, and losses through an informed and supportive IWC employees and a robust CAPP.

SEAP DRIVERS AND FLOWCHARTS

United States' court decisions have shown that, if a corporation does not adequately protect its assets *and* the employees do not know and understand their responsibilities relative to the protection of those assets, it is highly unlikely that a corporation will have a successful lawsuit against an employee for such things as theft of corporate property. If a complaint is accepted and the employee is prosecuted, the judge is likely to find that, if the employee was unaware of "the rules," how could the employee be expected to follow them? In addition, judges have ruled that if a corporation does not do a proper job of protecting its assets, it should not rely on the court to do it for them.

After gathering the information needed, one first has to look at security drivers, specifically the drivers for any corporation, in this case IWC, to protect assets. In other words, why is an SEATP needed in the first place? The security drivers identified are:

- Need to comply with federal, state, and local laws and government regulations.
- Comply with the laws of the nation-states where IWC does business.
- Comply with IWC internal policies, procedures, and directives.
- Comply with contractual requirements related to IWC assets used by others and the use of others' assets by IWC.
- Loss of IWC's valuable assets would adversely impact IWC's ability to successfully compete in the global marketplace.

Based on the drivers, the SEATP was developed.

SEAP METRICS

How does a CSO really know the impact of an SEAP on the company? Is the SEAP effective? To learn this, it will be necessary to measure the program in some way.

There must be measures put in place to track the costs, benefits, cause, and effects of an SEAP on the protection of IWC assets, as well as the time spent (potential productivity loss) by employees because they spend time attending training or awareness sessions that take them away from doing the "hands-on" job they were hired to do.

Every lost hour (time not spent producing goods or services of the corporation) by workers has an adverse impact on total productivity, and thus profits. If the SEAP adversely affects the productivity of workers, it must be able to demonstrate how it contributes to the protection of company assets, offsetting that loss of productivity—in essence, it must justify itself. SMMP techniques can assist in doing that.

One of the SEAP goals is to provide all briefings and training required to all applicable employees. The goal is, at the end of the year, to have reached 100% of the employees and support personnel. The effectiveness of training is measured by the correlation analysis of hours of training/employee at each site to all applicable assets protection policies and procedures at each location (measures effectiveness of training). The goal is a positive downward trend of the loss (theft, damage, destruction, etc.) of corporate assets. A reduction of security violations throughout the corporation is another potential indicator of success.

Each of the subfunctions and their products' development processes should be analyzed to determine their costs in terms of labor and materials. A useful tool to begin to accomplish this is—hopefully you guessed it—to flowchart the process.[1] In that way, each step in that process can be understood in the necessary detail to help determine its value. This will assist the CSO in developing an understanding of the real costs involved in the development of security products and assess the cost–benefits of each part of the SEAP. For example:

- What is the cost and benefit of developing and distributing awareness material such as posters displayed throughout the company containing a security message?
- Do the use of these and other SEAP products cause the employees to better protect the IWC assets?
- If not, what other purpose do they serve?
- Are they just a "nice-to-have" item with no visible return on investments? If so, eliminate them.

An SMMP will help the CSO make this determination. In times of corporate frugality, such products may be seen as too expensive and the CSO, when

[1]Remember all initial establishments of a functional or subfunctional SMMP should be through a project plan to ensure an organized, formal way of working this issue.

placing much reliance on them, may be looked upon as not understanding the business world of costs and profits. After all, a good business person must be able to demonstrate a return on investment. Using the tested measurement process, the CSO can demonstrate the value of the program. If the program has little value, the CSO must change or eliminate it. If the program demonstrated more of a benefit than it costs, the CSO should continue the program but also continue to strive to make it more effective and efficient.

Other considerations for the CSO include the effectiveness of the employee awareness briefings. Are they being accomplished cost-effectively? Are the security specialists conducting the briefings doing them for large audiences and thereby operating more cost-effectively by providing fewer briefings? By providing fewer briefings for more attendees per session, the security specialist frees up some of his or her time. That is time that can be used to perform other duties, contributing to the security program in other ways or making up for a shortage of security resources in other areas.

SAMPLE SEATP METRICS CHARTS

As with all security functions, there are numerous types and formats of data collection methods and chart development processes to support a CSO's assets protection analyses and decision-making processes. The following are a few examples of such SMMP-related charts.

Looking at the data, what questions might the CSO ask? Should there be concern about the uneven distribution of work between most months? Could there be a better way to deliver security awareness briefings? For instance, could there be a more efficient approach?

Graphic depictions of data should tell a story. They should also cause the viewer, in this case the IWC CSO, to see patterns, trends, or anomalies leading to questions about how this work can be accomplished more efficiently.

In this case, the briefings are conducted in order for IWC to come into compliance with federal law. There are many reasons to use the briefing process to increase employee awareness. In all cases there is a cost. However, that cost may be offset by adverse consequences. In this example, the consequence would be not complying with federal law. Consequences differ from situation to situation and requirement to requirement. How each is handled is very much related to how much risk management is willing to accept.

As can be shown by basic metrics charts, one of the primary metrics that should be used to manage an SEAP is the relationship to costs and productivity losses of employees attending the briefings, as well as the success of a CAPP or portion thereof.

It is important for the CSO to know how much it costs to develop and provide security awareness briefings, and what the impact is to company productivity, as well as assets protection. Furthermore, the CSO needs to be thinking about how effective those briefings are. Did employee awareness increase? If so, how do you know? One potential metric is to track the number of security violations committed and compare that to the number of persons briefed. If all company employees were briefed on their security responsibilities, did the number of security violations decrease or increase?

DATA COLLECTION AND METRICS MANAGEMENT

Some of the questions that may be asked by a CSO or a CSO's staff responsible for the SEAP are:

Who should track the functional metrics input data? As with all security functions, the tracking of the data should be done by the person at the lowest level who is responsible for the day-to-day SEATP activity.

What to track? To begin with, all major tasks should be tracked and, gradually, data collected in more detail as one moves through the work breakdown structure of the SEAP. To start, the tracking and data collection should be basic. The number of briefings and number of attendees at those briefings and the associated costs will provide a baseline of information to work with; the type, number and costs of all awareness material, and the costs of meeting all requirements (remember the security drivers) mandated action (e.g., laws and regulations) will also help establish an SMMP baseline.

Why track it? First and foremost, it is difficult at best to manage what you do not know. Therefore, tracking data and measuring performance is essential to know the basics about a process and what it does. Any process of high cost or frequency must be tracked as it is likely to get much attention. Once these factors can be quantified and qualified, then the process of analysis can begin to determine more cost-effective ways of providing the SEAP service and support to the CAPP and thus IWC.

How to track it? It is actually a simple process. To begin with, the tracking of costs and lost productivity of workers can be accomplished by maintaining a record of attendees (having a sign-in log for those attending the briefings) and the amount of time each spent at the presentation. This can then be entered into a spreadsheet and the total time spent at the briefing can be calculated, as well as the average time spent by each employee. This time can then be added to the average time it takes for an employee to get to the briefing and return to work. Multiply that time by the average employee pay rate plus benefits and you will have the total time of lost productivity due to attending SEAP briefings. Of course, the next thing to do would be to quantify the value of the briefings. This is a more difficult task. As the IWC CSO, how would you go about doing that?

When to track it? The tracking of various data from the processes making up the SEAP can be done at different intervals beginning with each occurrence and compiled—monthly, weekly, bi-monthly, etc.—and should be done in a way consistent with your organizational needs. Again, the criteria that should be used are how often the data is needed for analyses along with special actions and follow-up for CSO assignments and projects. To keep the process as simple as possible, data collection should occur as often as briefings are conducted and as often as awareness materials are distributed.

For distributed awareness material, the cost in time and other resources should be collected as awareness materials are developed. Added to that would be the publication and distribution cost, including material, labor, and other resources. So, for awareness material, the data would be collected at the end of each project. The total costs of all projects relative to the development of awareness materials would be summed on a monthly basis, again on a quarterly basis, and also on an annual basis; or, whatever your specific need happens to be.

Where to track it in the functional process? Each briefing's data would be collected at the end of each briefing and entered into the spreadsheet or database. Each awareness material development, production, and distribution would be collected at the end of the project for each awareness material project.

In analyzing these SEAP charts, the CSO can see if the ratio of attendees to briefings is cost-effective. Changes can be implemented on a trial basis to see if it has a positive impact, resulting in a more cost-effective briefing program. If so, excellent! If not, make another adjustment. The beauty of measurement is that it helps the CSO or security professional manage change, track results, and continue to make positive adjustments in the process working toward the most effective and efficient processes possible.

Another measure the CSO can use is the cost of lost productivity, because each employee must attend an annual briefing in person, therefore they are not spending that time being productive (in terms of their normal assignment and responsibilities). Assume an average wage of $50 an hour per IWC employee, which includes benefits as a part of that average wage. Then assume that it takes about 30 min for each employee to shut down their work and go to the briefing location and another 30 min to get back to the office and return to their primary duties. That means if you multiply the number of attendees by $50, you can estimate the cost factor of the employee's "travel" time alone. Now factor in another hour each at $50 an hour for listening to the 1 h briefing.

Suppose that you, as the CSO, learn employee security awareness must be achieved to a minimum standard in order to ensure all employees understand their responsibility to protect corporation assets. Essentially, your security awareness briefings are now required by federal law. What does this mean to you?

The law requiring that the employees be made aware of their duties and responsibilities for IWC assets protection may be interpreted (here you want to get a representative of the legal staff to provide guidance to you, the CSO) to mean that a briefing can be through a simple distribution of a security awareness pamphlet or an online briefing where employees can acknowledge their responsibilities as they go through charts and take a simple test to demonstrate they have learned the basic information. Which to choose? Keep in mind, your goal should be to determine which is most efficient and effective.

By placing the briefing online, there are advantages. It can be rapidly updated and tied to security software—used to collect data—so the relevant data can be collected in a centralized database for later metrics analyses. Furthermore, the travel time is eliminated so a savings of $50 per employee per briefing can be claimed as a savings and increase the employees' productivity.

There is of course the unknown factor and that is: By attending the briefings in any format, does that make the employees more aware of their assets protection responsibilities and are they then more apt to comply with the IWC CAPP and properly protect the IWC assets? If so, how do you know? One way to know is to look at the metrics of violations of the CAPP.

SEAP CASE STUDY

The IWC CSO met with the security department's investigations manager who stated that there was an ongoing security problem in IWC. That problem is the theft of IWC's information technology's assets used by employees, specifically theft of notebook computers out of the vehicles of IWC employees when the employees were out on business trips. It became particularly problematic when it was discovered that many of these stolen notebooks contained sensitive IWC information, information that was in many cases considered to be competitive-sensitive. That is to say, if IWC competitors were to obtain that information, IWC could see their competitive advantage negatively affected.

As the CSO, what would be your plan of action to eliminate the thefts and losses of these valuable IWC assets and the information they contained?

Of course, you should know by now that this calls for action managed through a project plan. In this case, the project plan consisted of the following tasks:

- Collect and analyze all investigative report data on the losses to include the usual who, how, where, when, why, and what, using a spreadsheet format.
- Coordinate the results with the SEAP security specialist.
- Develop an IWC-wide communication (it may be as simple as a corporate-wide e-mail notice) advising all employees of the problem and how to eliminate it.
- Update the new-hire and annual employees' CAPP briefings to emphasize the problem and solution.
- Update the CAPP policy to include a statement that all losses of notebooks and other valuable assets due to the employee being negligent in their duty to protect company assets would require them to pay for the loss (at least the equipment as its value is easily quantified) out of their salary. This will be an unpopular "pill to swallow" but it may get employee attention (of course, this must be coordinated with and approved by the legal and HR staffs). No employee would be authorized the use of a notebook computer or other assets, such as PDAs or cell phones, unless they signed a statement acknowledging their obligations to protect these valuable assets and having an understanding of the consequences. Failure to agree to such a practice will result in a restriction imposed on them to not remove any such assets from the facility.
- Develop and implement a useful security metric as a means of graphically depicting the future loss trends. Keep in mind that the goal is to drive down the number of losses and eliminate that asset protection problem.

Assume you are the CSO and are to analyze the charts and supporting data as well as provide a briefing to executive management personnel related to this problem. You must be prepared to clearly, convincingly, and concisely present your "story." Your graphic depiction of security metrics will support you in doing this.

> Always ensure that you have detailed charts and related documents to back up your overview charts and triple check them for accuracy, as any errors would detract from your reporting. Be sure you never present charts with inaccurate or inconsistent data. The integrity and quality of your charts are a reflection on you, the CSO. Bad charts equal a bad presentation with your message possibly being lost—or worse yet, not believed—at least in the minds of those subjected to your poor performance.

Any metrics trend chart should be followed by several other charts to include charts that show:

- Losses by items, e.g., notebooks, PDAs, cell phones
- Value of lost items
- Impact to IWC as a result of the item being lost or stolen
- Identify sources of losses, e.g., each department name

By depicting the losses by department, the CSO may generate a little competition between departments as they do their best to have the fewest losses in IWC. No one, particularly senior executives, wants to look bad in front of their peers or the CEO.

One way for a manager to eliminate the problem of lost high-value equipment may be to be sure that the losses are not reported (Gee, would a department manager actually not report a loss? Game the system? Yes!). However, this can be mitigated by regular assets inventories of certain types of high-value equipment.

SUMMARY

The SEAP is an important subfunction under the administrative security function. It is primarily responsible for awareness briefings, development and distribution of awareness material, and also for administratively managing the security departments' security professionals' training program.

The SEAP costs in terms of lost productivity and the loss of other resources can be measured and should be measured using a security metrics management approach. In addition, it can be used to help eliminate a trend of assets losses by emphasizing the problem and solution through awareness materials and briefings monitored through metrics trend charts in conjunction with other protective measures.

Security metrics charts, such as the ratio of briefings to employees attending, lost productivity caused by attending briefings, costs and benefits of awareness materials, and cost saving in terms of identifying assets protection loss trends and mitigating those losses through enhanced awareness materials and briefings, are the basis for an SEAP supported by an SMMP.

CHAPTER 7

Surveys and Risk Management

CONTENTS

INTRODUCTION

The term "surveys" as used here basically refers to a combination of risk management methodologies combined with targeted, proactive evaluations of various groups of valuable assets to determine how well they are protected, what the threats are against them, their specific vulnerabilities, any risks to them, the protection measures in place, and their costs, as well as identifying more cost-effective methods for the assets' protection.

The security surveys are an exclusive security function to be performed only by the administrative security organization's security professionals. As needed, support will be drawn from other security functions or external experts. Within the International Widget Corporation (IWC), these security surveys are best performed within the administrative security organization.

The risk management security function is an integral part of the security survey process. However, the Chief Security Officer (CSO) has mandated that the risk management philosophy and related methodologies be part of the decision-making process of the entire security department's staff.

Security surveys are often used to target certain assets for evaluation, with emphasis not on compliance with the corporate assets protection program (CAPP) but from the viewpoint of risk reduction. Essentially the survey helps the CSO better protect an asset from threats while recognizing and considering its vulnerabilities.

For example, a security survey may look at the threat agent potential to dial-up computer systems within IWC. That may include using a software program to dial all the telephone numbers used by IWC and document those that acknowledge with a modem tone. Penetration attempts would be made against those systems using common hacker tools. The findings would be documented and

Security Metrics Management. DOI: http://dx.doi.org/10.1016/B978-0-12-804453-7.00007-0

reported to IWC management along with recommendations for protection changes, revision of the CAPP, and other applicable changes.

Another security survey may be to use social engineering techniques to try to gain information from employees and possibly gain physical access to facilities.

Security surveys are a more proactive way of determining if assets are properly protected, as the security professional conducting the security surveys uses common threat agents' techniques to try to obtain access to valuable assets.

The risk management methodologies are then applied as part of that approach. Risk management is an often misused term and is a philosophy and methodology that is sometimes incorrectly applied.

> As part of any executive management briefing, the CSO must provide information as to the overall risk management strategy that is to be integrated into the security department staff's assets protection decision-making process and the reason for that approach.

When it comes to risk assessments, risk analyses, and risk management, there are those who argue quantitative versus qualitative. Regardless of your preference, risk assessments are still only best guesses—ideally, a best educated guess.

SURVEYS AND RISK MANAGEMENT DRIVERS AND FLOWCHARTS

As with all security functions, the security surveys security metrics management system requires that we begin with the drivers and the flow process that follows.

THE WHO, HOW, WHERE, WHEN, WHY, AND WHAT OF SURVEYS AND RISK MANAGEMENT METRICS TRACKING

As we have previously stated in other chapters, the first step in data analysis and measurement is data collection. With each security function and its processes, the person(s) responsible for tracking and collection must be identified. Moreover, what, when, who, where, why, and how of data collection, measurement, analysis, and depiction must also be identified.

Who should track the functional metrics input data? As with all security functions, the tracking of the data and data collection should be done by the person at the lowest level who is responsible for the day-to-day leadership role for the security surveys.

What to track? To begin with, all major tasks should be tracked and, gradually, data collected in more detail as one moves through the work breakdown structure of the security survey function. For this function, the tracking and data collection should be that of the number of surveys conducted, their location, any departments who are found to be deficient, the survey results, their impact on assets protection and, of course, all of their associated costs; the type and frequency of all resources involved; and in particular, the costs of meeting the drivers' requirements, in this case the primary driver, which is the potential threats, vulnerabilities, and risks to the specific assets being surveyed.

Why track it? This function should be tracked because it is a function that is vulnerable to elimination when cost-cutting time comes around. This is a valuable function but, unlike most of the others, this one could be eliminated or integrated into various other security functions as part of the other functions' processes. Security surveys have little to do with accomplishing the routine tasks a security organization faces each day. They have much to do with assessing how well processes are working. During periods of high budgetary pressure, processes that do not directly contribute to the day-to-day security operation are often eliminated or deferred. Therefore, it is important to continually identify the cost–benefits of this function and its associated return on investments.

How to track it? It should be tracked through each step and as part of a subprocess in the individual security survey's operations plan.

When to track it? The tracking of various aspects of the security surveys should be done by updating all related spreadsheets and databases as each security survey is completed.

Where to track it in the functional process? Each survey's data would be collected as each step in the operations plan was completed. The collection point would be at all the primary points noted in the process flowcharts for the security surveys, as well as during each contact with an employee and its duration with a member of any department being impacted by the security survey.

As with the security compliance audits and other security functions, analyzing the security survey metrics charts, the CSO can begin to understand the costs and impact, both positive and negative, of the security surveys.

The charts also can be used to look at the costs in lost productivity as would be done with all the other security functional metrics management systems.

CASE STUDY

With the increased threats to IWC assets throughout the world as IWC expanded on a worldwide basis, IWC executives wanted to know what the

IWC CSO is doing to understand and mitigate today's threats to corporate assets and, in particular, the potential threats of terrorism.

The CSO thought that the best way to provide that information was with a basic briefing addressing the potential threats related to IWC worldwide, how those threats were viewed by the security staff, and what was being done to mitigate the threats.

The CSO viewed this matter as too important to be glossed over lightly and decided an introductory briefing was best suited for executive management, with follow-on briefings to be scheduled as the executives deemed appropriate.

The CSO provided the following information to the IWC executive management in the form of an executive management briefing and discussion.

Introduction

- This briefing will discuss the general characteristics of threat agents that can be identified and measured as an integral part of a security survey incorporating risk management techniques with emphasis on cost–benefits and return on investments.
- The method has been tested against historical data (in this case with open source information) to establish the validity of the metrics that have been developed.
- Other characteristics of threats are outside the scope of this presentation.

Why Develop a Threat Method?

- The measures that are presently used to protect corporate assets need to be cost-effective.
- Returns on investments must be identified.
- Current risk assessments rely on knowledge of the probable threat(s).
- The need to develop a system that allows scenarios to be modeled.
- Need to produce a threat assessment that can be understood and used by management in their decision-making processes.

The Threat Problem

- Current practice for the production of threat assessments is the use of "experts" who have considerable experience preparing assessments from all sources of intelligence.
- The whole process is currently subjective and cannot be accurately modeled.

- The production of a threat assessment cannot be easily repeated at short notice (scarce resources, tasking of intelligence assets, time taken to produce assessment).
- The logic and the factors that were used in developing the assessment cannot be easily tested.
- The resources best used to produce threat assessments are not normally available for business and industry and are not in sufficient detail to meet the needs of organizations comprising the Critical National Infrastructure.
- Most of the available "experts" have little or no knowledge of the information environment—their expertise was gained in the physical and personnel areas.
- At IWC, understanding the information environment is crucial due to IWC's reliance on high technology as the underpinning of our competitive edge.
- Based on these issues, surveys with integrated risk management techniques are used here at IWC.

Preliminary Research

- Before work started on developing a method for threat assessment, the following areas were examined for any relevant concepts or methods:
  - Risk assessment methods
  - Insurance underwriting
  - Gambling (horse and dog racing)
  - Government methods (the FBI Computer Crime Adversarial Matrix)
  - Construction industry
  - Airline industry
- This was done in order to adopt and adapt validated methodologies already in place.

The CSO went on to explain the various aspects of the threat assessment philosophy as part of the surveys and risk management (SRM).

The CSO went on to explain that a values matrix can be developed and some calculations made that can be used to quantify the risks of today's threat agents to IWC assets.

The CSO told the group that once values have been established for each of the threat elements, it is possible to establish an overall value for the threat. This is not an absolute and will currently only give a relative value. It will allow for the threat from various agents to be compared and it will allow a range of scenarios to be modeled and compared.

When conducting any type of SRM relative to potential terrorist threats to IWC, the CSO's SRM team also considers other issues such as:

- Likelihood of an attack
- Likelihood of a successful attack
- Value of lost revenue
- Cost of system repair
- Cost of reputation damage
- Third-party damages
- Cost of loss of confidence

The CSO continued to explain that there are current and foreseeable threat scenarios not only from terrorists but also:

- Pressure groups
- Collateral damage from other events
- Organized crime
- Nation-state government threats where IWC has offices

Using a security survey approach, the IWC CSO explained that the security department will maintain current information relative to terrorists and other miscreant's attack techniques and test them (obviously in a simulated or controlled manner) on a periodic basis.

A security metrics management process will be put in place to collect data for risk analyzed by the security survey personnel. The data would be compiled into charts for the CSO's monitoring of the threat-related risk management issues. They would also be used to support periodic reports and briefings to the IWC executive management.

The data collection would include:

- number of security surveys conducted relative to terrorism,
- the targeted areas,
- the results,
- the costs of the surveys,
- the value of the assets currently and adequately protected and at what costs relative exclusively to the terrorists threats,
- the protection deficiencies identified through the surveys,
- the risks to those assets,
- the cost of mitigating the risks,
- status of implementation of additional protective measures.

SUMMARY

Conducting security surveys with an integrated risk management approach, when done in a proactive way, can help provide more cost-effective and successful assets protection programs.

The metrics that are to be used in this process should include data collection of the time spent conducting the surveys and their results in terms of acceptable levels of risks regarding assets protection.

Of particular interest are the threats posed by terrorist groups that may target IWC due to the location of its international offices (Asia, Africa, Middle East) and also since it is a US-based corporation.

CHAPTER 8

Contingency Planning

INTRODUCTION

Within corporations, security is often the organization responsible for contingency planning. This responsibility is usually attained as a default action. Since the security organization is the organization serving as the primary first responder to an incident or emergency, the transition to assuming the lead responsibility for contingency planning is a logical one. It is the experience as a first responder that leads to an expansion of that role and into the role of the contingency planner.

Since the World Trade Center attacks, corporations and government agencies have become more sensitive to the need to plan and prepare for an emergency or crisis. To that end, security professionals in the United States and some other countries have expanded their contingency-planning capabilities beyond their traditional responsibilities for emergency response and crisis management to focus on the effort to recover from such an event. It is becoming quite common to see security professionals responsible for the entire spectrum of contingency planning. This includes the following elements:

- Emergency Response
- Crisis Management
- Business Continuity—consisting of the following subelements:
  - Business Recovery
  - Business Resumption

It is critical that security professionals learn from actual events and planned exercises (drills) and apply that learning to improve future actions. Contingency planning is and must be a dynamic process where systems are tested and regularly revised. The security professional cannot afford the all-too-common situation of developing comprehensive plans but never testing them or never changing them as they learn from actual events. In simpler

CONTENTS

Security Metrics Management. DOI: http://dx.doi.org/10.1016/B978-0-12-804453-7.00008-2

words, contingency plans do not belong on shelves collecting dust. They must be read, revised, tested, and shared to be as effective as possible. Keep in mind that perhaps more important than the planning documentation itself is the process of learning what occurs by everyone who participates in the planning, testing, and implementation of such plans.

A Chief Security Officer (CSO), such as the International Widget Corporation (IWC) CSO, who is responsible for contingency planning, must have a clear understanding of what this all means. After all, much rests on the CSO's ability to do this well.

Since contingency planning is as much a learning process as it is a planning process, it is a good idea to start with a basic foundation. The most important part of that foundation is establishing a common language. Many terms are used in the process of contingency planning. It is critical that all parties have, and understand, a common set of definitions. Below are samples of definitions for some key terms used in the contingency planning process.

If these definitions do not fit well within your organization, change them. What is critical about these definitions is not that they be standard throughout all of business and industry but that, when they are used, they have a common meaning understood by everyone within the organization using them. They can be tailored uniquely to your own organization's needs.

- *Business Continuity*: Minimizing business interruption or disruption caused by different contingencies. Keeping the business going. Business continuity plans encompass actions related to how one prepares for, manages, recovers, and ultimately resumes business after a disruption.
- *Business Recovery*: Refers to the short-term (generally <60 days) restoration activities that return the business or government agency to a minimum acceptable level of operation or production following a work disruption. Used interchangeably with the term *disaster recovery*.
- *Business Resumption*: The long-term (generally >60 days) process of restoration activities after an emergency or disaster that return the business to its pre-event condition. (Keep in mind that restoration to the exact pre-event condition may not be necessary or even desirable. However, making this determination may not be possible without proper planning or going through the actual resumption process.)
- *Contingency*: An event that is a possible but uncertain occurrence or is likely to happen as an adjunct to other events.
- *Contingency Planning*: The process of planning for response, recovery and resumption activities for the infrastructure, critical processes, and other elements of IWC based upon encountering different contingencies.

- *Crisis Management*: The process of managing the events of a crisis to a condition of stability. This task is best accomplished by one of the businesses' integrated process teams (IPTs) made up of members from different disciplines throughout the company. This IPT serves as the site or business deliberative body for emergency response and crisis management planning and implementation.
- *Critical Processes*: Activities performed by departments which, if significantly disrupted, due to a major emergency or disaster, would have an adverse impact on the business operations, revenue generation, customer schedules, contractual commitments, or legal obligations.
- *Emergency Response*: The act of reporting and responding to any emergency or major disruption of a business.

  The process of contingency planning at IWC is focused to achieve the following:
- *Secure and protect people*—in the event of a crisis, people must be protected.
- *Secure the continuity of the core elements of the business*—the infrastructure and critical processes—minimize disruptions to the business.
- *Secure all information systems*—that include or affect suppliers' connections and customer relationships.

CONTINGENCY PLANNING DRIVERS AND FLOWCHARTS

The main driver for contingency planning is of course the need to protect corporate assets and that need derived from the other drivers stated earlier, such as laws and regulations. That need must include the ability to rapidly get back to normal business in the event of an emergency, disaster, or other significant business disruption.

Contingency plans formally establish the processes and procedures to protect employees, core business elements, information systems, and the environment in the event of an emergency, business disruption, or disaster. These IWC plans, also incorporated into the IWC corporate assets protection program (CAPP), discuss specific types of emergencies and disasters and address the mitigation, preparedness, and response actions to be taken by IWC employees, management, and the organizations charged with specific response and recovery tasks.

These plans contain basic guidance, direction, responsibilities, and administrative information. At IWC, the CSO recognized the criticality of regularly evaluating and improving company contingency plans. To that end, the CSO formed a project team to evaluate the entire IWC contingency program. Furthermore, it was decided that effective measures must be put in place to assist in managing the successes, failures, and costs of the program. The IWC CSO needed to develop a metrics management program. Before developing this program, the IWC CSO needed to further define the parameters of the company contingency planning program. The following section addresses those parameters.

The project team decided that to develop contingency plans, the preparedness process must include the following considerations and elements and be supported by a security metrics management program (SMMP):

- *Assumptions*: Basic assumptions need to be developed in order to establish contingency planning ground rules. It is best to use as a baseline for planning, several possible "worst case" scenarios relative to time of event, type of event, available resources, building occupancy, evacuation of personnel, personnel stranded on site, and environmental factors such as weather conditions and temperature. Furthermore, consideration should be given to establishing response parameters for emergency events. Define what constitutes a minor emergency, a major emergency and a disaster.
- *Risk assessment and vulnerability analysis*: An IWC crisis management team was recommended by the project team and subsequently formed, with the responsibility to identify known and apparent vulnerabilities and risks associated with the type of business and geographic location of the enterprise. An assessment of risk and vulnerabilities will be made prior to upgrading contingency plans. All planning will be accomplished in accordance with a thorough understanding of actual and potential risks and vulnerabilities. The project team's risk assessment and vulnerability analysis also included an assessment of the policies and practices of IWC's critical relationships. That meant involving suppliers and customers in the contingency planning process. Regardless of how prepared IWC may be, if a critical supplier or many key suppliers are not also prepared for various potential contingencies, their inability to recover will adversely impact IWC. Therefore, critical suppliers will be integrated into the contingency planning process.
- *Types of hazards*: Planning for each and every type of hazard is not practical nor desirable. Grouping hazards into similar or like categories will allow for planning to address categories of hazards. Since many hazards have similar consequences and result in like damages, it is best to plan for them in categories. The following is a list of common hazards IWC may face:
  - Medical emergencies
  - Fires
  - Bomb threats
  - High winds

  - Power interruptions
  - Floods
  - Hurricanes/typhoons
  - Snow/ice storms/blizzards
  - Hazard materials issues
  - Aircraft crashes
  - Civil disorders
  - Earthquakes
  - Terrorist threats/activities
  - Workplace violence
  - Explosions
  - Tornados
- *Critical process identification*: The project team also decided that all critical processes must be identified. These processes must be ranked in accordance of criticality and importance to the productivity and survivability of the enterprise. The process of recovery will be focused on those critical processes that, when resumed, will restore operations to a minimal acceptable level. In essence, these processes are identified to be the first processes restored in the event of a major interruption to business operations. Failure to restore them presents the greatest possibility of damage or loss, including possible loss of IWC's competitive edge and market share.
- *Business impact analysis*: A business impact analysis must be accomplished to accurately determine the financial and operational impact that could result from an interruption to the IWC business processes. Moreover, all critical interdependencies, those processes or activities which critical processes are dependent upon, must be assessed to determine the extent they must be part of the contingency planning process.
- *Emergency response*: Establishing precisely who will respond to emergencies and what response capabilities are needed was considered by the project team to be essential. All participants in the emergency response process must understand what is expected of them. These expectations must be well defined and documented. Guidance for all employees on how to react in the event of an emergency and what their individual and collective responsibilities are must be documented and distributed. Organizational responsibilities must also be established to include the development of department-level emergency plans. Events such as building evacuation and roll-call assembly need to be well defined so in the event of an actual emergency, there is no confusion or uncertainty as to what must be accomplished.
- *Incident management and crisis management*: The project team determined that as an incident escalates, the crisis management team should

assume the responsibility of managing the crisis. How this process works and who has what responsibilities must be clearly stated in the contingency plans. In the event of an actual emergency, there will be people who will attempt to manage the incident or participate in crisis management; however, they should not have a role whatsoever in this process unless they were previously identified and trained as part of the crisis management team. Without established and well-defined incident management protocols and procedures, chaos is likely to occur.

- *Incident/event analysis*: When an emergency incident or event occurs interrupting or disrupting the IWC business process, the IWC security department personnel will be charged with responding to and managing the scene. They will also be responsible for conducting an incident/event analysis. This analysis will be conducted to determine the immediate extent of damage and the potential for subsequent additional damage. The appropriate resources must be notified and activated to assist in damage mitigation.
- *Business resumption planning*: The project team decided that the process of planning to facilitate the recovery of designated critical processes and the resumption of business in the event of an interruption to the business process must be performed in two parts. The first part focuses on business recovery in the short term while the other part focuses on business restoration in the long term. This process will also include establishment of priorities for restoration of critical processes, infrastructure, and information systems.
- *Postevent evaluation*: An assessment of preceding events to determine what went well, what went worse than planned, and what improvements to existing plans should be made is also part of the process. Learning from real events is an unfortunate opportunity. There is no better way to learn how to handle an emergency than to actually handle one. Unfortunately, experiencing an emergency may cause damage to IWC.

Now that you are familiar with the SMMP methodology, identify the various data, its purposes, and describe charts that would be used by an SMMP to support this effort.

As part of the contingency planning effort, the CSO wanted to develop a graphic depiction of the contingency planning function that shows how the business continuity planning program should be structured to include depicting the major elements of that structure. This tool could be used to help the CSO and corporate management focus on identifying areas where process measurements could provide a valuable insight into the effectiveness or efficiency of critical components of the contingency planning program. Additionally, the graphic depiction could be used as a training tool with executive management and all participants in this vital function.

Another tool the IWC CSO decided would be useful would be a graphic depiction showing the relationship between IWC's crisis management team and business continuity team. As with the previous chart, the CSO will use this chart as a tool to assist in identifying potential metrics points and for briefing and educating executive management.

At IWC, the information technology department and the information systems security (InfoSec) function were very closely related. In addition, the widget manufacturing was based entirely on computer systems and their associated networks. Therefore, the IWC CSO wanted a flowchart developed that would show the relationship between the general security functions in the event of an emergency and the InfoSec function led by the information systems security officer (ISSO).

Working with the CSO-appointed contingency planning project team, the ISSO and team developed a flowchart that depicted how they were currently established to deal with an emergency.

The CSO decided that this chart was sufficient as a tool to be used to demonstrate how communications and functional processes worked between contingency planning leader (CSO) and the information systems security lead (ISSO). It was also a tool that could be used to help identify critical areas in the process requiring regular measurement. Supporting this effort, process flowcharts could be developed to further break down the process activities.

In furtherance of understanding the contingency planning process, the IWC CSO determined it necessary to develop a threat matrix to link together the relationship between specific assets and infrastructure and related threats. The intent was to focus limited resources on planning for categories of scenarios that have a high likelihood of occurrence.

The matrix chart can be used as a tool not only to determine if the threats are mitigated through contingency planning policies, procedures, and processes but also to test various aspects of the contingency plan by targeting potential threats to specific assets.

EXAMPLES OF CONTINGENCY PLANNING METRICS' MEASUREMENT TOOLS

One of the crucial elements of preparing for and especially reacting to emergencies is time. How quickly can an emergency response team, crisis management team, or business recovery team react to any crisis? Since the precise time and location of a crisis or disaster cannot easily be predicted, preparation is difficult at best. For example, the CSO can anticipate a natural disaster will occur at his site in Florida during the hurricane season when a hurricane is identified as bearing down at the IWC location there. However, the CSO does not know if or when a large earthquake would occur in California. The best way to ensure contingency planning is effective is to test the plans before having to implement them during a real event. Ideally, the CSO and corporation will learn through testing and not the real event.

Much can be learned through testing (conducting drills). Gaps in the plans can be identified, team members will learn and become better prepared to react through practice, supporting resources can be identified, and reaction times can be simulated. Does the CSO have any idea as to how long it will take to respond to a disaster and implement the appropriate contingency plan? The testing process will help the CSO learn and improve.

Reaction and implementation time, along with other related contingency planning questions must be asked and answered through the testing process. The answers will help the CSO better frame the overall plan in a manner that considers the areas of highest risks and greatest vulnerability. With limited resources available, directing them at the most critical of processes is essential.

To address the reaction time problem, the IWC CSO directed that under the SMMP time and motion studies would be conducted to test each element of the contingency planning program to ensure that the processes implemented are effective and efficient. By that, we mean all planned actions are taken that would be necessary to maximize the mitigation of the threat.

The CSO's contingency planning project team developed a "Time–Motion Study Collection Data Sheet" to be used when testing various aspects of the contingency plan. Again, testing such a large-scale plan cannot be done at once, as this would neither be an effective nor efficient way to measure the results. As the saying goes, "How do you eat an elephant? A bite at a time!"

All components and processes of the contingency plan can eventually be documented and captured in a database as part of an information processing system for analyses. The value of resources necessary to prepare and implement contingency plans can be established and compared with the potential losses to IWC should a disaster strike. Furthermore, through drilling, redundant or unnecessary actions can be eliminated, thereby streamlining the process better and enabling the team to react and implement more efficiently. All these changes should be documented to include time variances between the old process and the new process, as well as materials used that can be translated into costs and incorporated into the SMMP.

Responding to an emergency requires skilled and trained persons working to a "practiced" plan. Recovering from a disaster quickly so as to minimize losses incurred with the operation shutdown requires meticulous planning, preparation, and testing.

As a CSO, are you prepared to handle a crisis as effectively and efficiently as possible? What actions can you take to help you get there if you are not? You can start by identifying in detail the contingency planning function and its components. How do the many processes work? Once that is established, test each process. In some case only a "table top" test will be necessary or practical (such as preparing to obtain additional monetary assets from a lender to assist with a situation). In other cases, an actual test will work (such as testing how long it takes to get a response team from one point to another). Through process analysis, time/motion studies, and conducting test/drills, much can be learned to "perfect" the process. Document the results into the SMMP.

Keep in mind, in each phase of analysis, study and testing, data is collected and measurements are made. From this point, efficiency and effectiveness gains can be identified through additional data collection tools with the goal to continue to gain efficiencies while increasing effectiveness.

Once this portion was tested over time, an average response and recovery time can be established for each tested threat matched to each asset. This will take time to accomplish but, in both the short term and the long term, the CSO can begin to more effectively and efficiently manage this function.

CONTINGENCY PLANNING CASE STUDY

The Chief Executive Officer was concerned about IWC being prepared for any emergencies and disasters. The CSO was asked to provide a briefing to the executive management to inform them as to how well IWC was prepared to handle such incidents. As the CSO, what information would you provide the executive management using what metrics-related charts?

Much of what you brief and how you communicate it will be based on the culture and working environment at your corporation or government agency. At IWC, the CSO provided the following metrics-related briefing charts:

- Contingency planning internal and external (government regulations) drivers
- How the contingency plan was integrated into the total IWC CAPP
- How IWC was structured to react in the event of an emergency or a disaster
- An example of the process flow if an emergency was declared

- An example of the process flow if a disaster was declared
- An example of how recovery would take place
- What IWC personnel were doing to prepare for emergencies and disasters
- The costs of preparation
- Examples of estimated costs of recovery using several emergency and disaster preparedness charts
- Results of the tests conducted year-to-date broken down from a summary chart to various types of tests conducted within what departments, etc.
- An overall summary chart

SUMMARY

Contingency planning may not be a traditional security process but, in today's global business environment, corporate security is assuming a much greater role and responsibility for its implementation. Even prior to the events of September 11, 2001, many organizations were becoming more conscious of the need to have contingency plans. September 11 accelerated the process for many, including IWC. A complete contingency planning program has three major elements:

1. Emergency response
2. Crisis management
3. Business continuity: business recovery and business resumption

Contingency planning-related functions require complex preparation, documentation, and testing. Such a critical business process—the contingency planning process—must be made as effective and efficient as possible. Measuring that process through process analysis and testing using metric tools designed to meet your organization's needs will ensure the overall process is as effective as it can be. Security metrics management can help manage this important function in a cost-effective way.

CHAPTER 9

The Guard Force

INTRODUCTION

CONTENTS

Physical security is the first line of defense in the protection of corporate assets. Physical security has been a part of assets protection longer than any other security function and it is also often the most expensive. Security guards, the "human part" of the physical security functions, have probably been around the longest. Ever since there was something of value that owners wanted protected but could not do the job themselves, security guards have been employed to protect those owners' assets, as well as the owners themselves.

Security guard forces are either proprietary—in other words, they are directly employed by the company or government agency they support—or they are outsourced and belong to a security services provider. Proprietary security guards are employees of the company. Outsourced security guards are employees of a different company—one that sells security services. Security guards serve many roles as part of the physical layer of protection provided for a company.

Those roles include:

- Controlling the physical access of personnel to the sites, facilities, or areas they protect.
- Controlling internal access to sensitive areas inside the buildings of the facilities they protect.
- External roving patrols on the corporate property, where they serve as observers and deterrents.
- Internal roving patrols inside the corporation's property as both observers and deterrents. They may have many additional tasks assigned to them in this role but observation and deterrence are usually the primary ones.
- Control center operators, which usually include alarm monitoring and closed-circuit television (CCTV) monitoring.

Security Metrics Management. DOI: http://dx.doi.org/10.1016/B978-0-12-804453-7.00009-4

- First responders to incidents and emergencies.
- Support the emergency response and crisis management process.

As part of the "front line" of protection and first responders, security guards are often the first on the scene to provide aid to the sick or injured and are therefore almost always trained in basic first aid. They also are usually the first organization representative a nonemployee sees when visiting a corporate office. In this capacity they serve as one of the corporation's unofficial public relations personnel. They are the initial image of the corporation. Because they make regular contact with most if not all corporate employees, they also become somewhat of an informal information officer.

A good use of a security metrics program will include data collection relative to all the work the security guard force performs. That, coupled with some time and motion studies, will help determine if the security guards' time is being used cost-effectively in the protection of the corporate assets.

For example, more than once a roving security guard walking the inner areas of a company has been asked by someone who is "very busy" to do them a favor. When the guard passes by the company cafeteria, could the guard please bring back a sandwich for the "too busy" employee? The guard, wanting to be nice, obliges and soon it becomes more of a habit than a one-time favor. As a Chief Security Officer (CSO), you are paying for this "room service."

One of the difficult decisions that must be made is whether or not to arm security guards. Of course, the main consideration with this decision is one of corporate liabilities. Is risk to corporate assets at such a high level that it warrants the use of armed security guards? Clearly in many instances it is and also in many instances, it is not so clear. However, with the ever-increasing threats of terrorist attacks against corporations, especially at corporations operating in hostile locations throughout the world, the need to arm security guards should be first and foremost based on the perceived risks to assets, as well as the value of those assets. There may also be liability issues for not adequately protecting assets—keep in mind that includes employees—when dangers are known or when risks are high. One must also consider whether the local government allows the arming of corporate security guards.

As today's and tomorrow's advanced-technology physical security-based protection devices continue to become more capable and less costly, functions traditionally performed by security guards are gradually being performed with high technology—or at least the support of high technology.

From CCTV cameras with smart software capable of recognizing unusual behavior or situations, to the use of biometrics and other sophisticated sensors, technology in many cases can do much more than a human. However, as long as assets require physical security protection, there will always be a need for a human to respond to the scene of alarms and emergencies, at least until the day arrives when sophisticated robots can take their place. Furthermore, there are situations where human judgment is needed and that has yet to be effectively replicated through the use of high technology.

GUARD FORCE SECURITY DRIVERS AND FLOWCHARTS

As with all security functions that are related to assets protection, there are certain drivers that cause the need for security guards:

- Laws and regulations
- Corporate policies
- Risk management decisions
- Executive management decisions
- Contractual requirements that impact physical assets protection
- Input of employees relative to physical assets protection
- Situational events and emergency situations

These all drive the need for physical security. A major portion of most physical security protective measures includes the use of security guards. Understanding how effective they are as part of the protective profile and the corporate assets protection program (CAPP) is the job of the CSO.

As with all security functions, an analysis of the function should be made and graphic depictions developed of how the processes work. That includes detailed flowcharts relative to the security guard function, procedures, and processes.

The guard force flowchart shows the basic process used for IWC's proprietary guards. This chart is then the basis for more detailed flowcharts. However, even from this initial flowchart, one can see an area where the security metrics management process might help reduce costs. Can you identify that particular part of the flowchart? (See the case study below.) Often, the easiest way to begin the security metrics management program (SMMP) is to begin with the first step of a security function as shown in their flowchart. After that is in place, move on to the next step and the next, etc.

GUARD FORCE METRICS CHARTS EXAMPLES

As we stated earlier, security guard forces are probably a corporation's most expensive form of physical security. Therefore, it behooves the CSO to track the costs of the security guard force and an SMMP is one of the ways of doing so. It does not matter if the security guard force is proprietary or managed as an outsourced contract. It will still be expensive, and therefore the costs–benefits should be tracked and analyzed.

There are certain security processes that are not directly related to the protection of assets. These processes should be addressed first because, as a CSO, you are paying for security guards that are to protect assets; however, they are not always doing that. For example, consider the hours each guard spends in meetings and the total number of hours and costs of security guards spent in preposition posting meetings. Yes, providing information and communicating with the shift's guard force is necessary; however, there may be better ways to do it.

One obvious objective is to find ways to either reduce such meetings or some way to eliminate them altogether. Quite often, security guards are being paid overtime, which can be expensive. One security metric would be to track the scheduled work hours against the overtime hours and then find the systemic cause of the overtime hours. The next step would be to try to eliminate or at least minimize the overtime hours. For example, some security guard forces are paid from the time they show up at work, punch-in, and then sit around awaiting roll-call. Obviously, this would be one area to try to eliminate. Although one must always be cautious—guards are people, too—in working these projects and reducing costs, while trying not to adversely impact guard force morale. That will not be easy, especially if you want to take away the overtime and impact their expected compensation.

One last word of caution: Some proprietary guard forces are unionized. That means that any changes must be negotiated with the union. Current contractual agreements may preclude immediate action; however, the issue must be considered when negotiating a new contract.

When dealing with a union guard force, the CSO must have a good understanding of the union contract and all related issues. Failing to do so may adversely impact the CSO's ability to find a way to achieve cost efficiencies. If a CSO does not understand the contract and how to go about using an SMMP with a unionized security guard force, the CSO does so at his/her peril. A CSO would be smart to work with the human relations specialists in this area along with the company legal staff and contract specialists in order to stay out of trouble, as security guard force processes and procedural changes are always a very sensitive issue.

GUARD FORCE CASE STUDY

The guard force at IWC is a proprietary guard force—all security guards are employees of the IWC. The guard force flowchart should depict the various functions and processes at a high level. As you develop your security functional and process flowcharts, you would of course go into much greater detail. For example, you may drill down into subprocesses, depicting such processes as the assignment of vehicles and the process for maintaining them.

However, even this high-level chart provides some points where metrics measurement can and should take place. For example, there is that preposting guard meeting. The guard force is made up of hourly employees. That means that they must attend the pre–post meeting. Since they are hourly employees, they must be paid for this time. The "problem" is that they are paid to be on duty protecting IWC assets. They are not doing that sitting in a meeting. Yes, the meeting is important, but what is the cost of that meeting? Is there a more efficient yet effective way to meet the objectives of that meeting? Looking at it from an SMMP perspective may provide some ideas once the data is collected.

Let us assume that at IWC the proprietary guard force is a nonunionized guard force. Therefore, there are fewer complications because there are no contract issues to deal with. Now, how would you begin to look at efficiency gains, thus making the guards more productive? This would include:

- Being on duty on time.
- Providing them the required information.
- Ensuring that they are presentable in appearance.
- Save money by not having them attend the preposting meetings.

Of course, there are some things that quickly come to mind, such as having computers built-in to the patrol vehicles, LAN connections, or mainframe dumb terminal connections at each post. They would receive up-to-date information via their computer terminal and it would also allow them to sign-in for work by logging into the system. In addition, their assignments for the next day would be sent to them via e-mail before the end of their shifts. That way they could just show up the next day at their posts for work.

As far as proper appearance, one must trust them to have sufficient pride in themselves that they would show up for work in proper appearance and in proper physical condition. In addition, the shift supervisor must make the post rounds and could determine if their appearance met IWC standards.

All of that sounds good on the surface, but what are those costs? So, you see, we are getting ahead of ourselves. First, we should look at this project by isolating the target to be the pre–post meeting. That being the case, what is the first question that must be asked?

Hopefully, you immediately know that answer: "What are the drivers that require that meeting to take place?" Yes, every stage of the process has drivers that drive the process to be performed the way it is being performed.

However, who decided that was the most effective and efficient way of meeting the objectives that were to be met by the pre–post meeting? Was it decided after a thorough and detailed analysis? Probably not.

It was probably done that way because that is the way it was done in the military or law enforcement, and the person who was hired to manage the guard force may have been ex-military or ex-law enforcement—or maybe retired from one or the other. Also, there is the old fallback reason: "We have always done it this way!" (Yes, that way since at least the last two millenniums!) The point being, you can pretty much assume very little formal analysis was done on which to base the decision to have a meeting.

The objective of the metrics management approach to this project is to save time and therefore money by finding more efficient and yet at least as effective a process to meet the objective now being met by the pre–post meeting.

So, what are the steps, the tasks, to be completed for this effort? They should include:

- Identify the drivers that make the meeting necessary.
- Validate the drivers as actually being necessary.
- Develop a *detailed* process flowchart of the pre/post meeting process.
- Analyze the process flowchart to identify centers of data collection.
- Develop security metrics management spreadsheets and charts that are to be used to graphically and statistically display for analyses and briefing the results of the security metrics management process.

The charts shown in the appendix are to be used by the IWC guard force manager and the CSO to support at least one of the project goals, and that is the goal of determining the true cost in terms of hours that can simply be converted to a monetary value. This is an efficiency gain goal and not a productivity enhancement goal since the guards are being paid to attend a meeting and not protect assets at the meeting.

Let us look at an example: If you have a 20-person-per-shift guard force and shifts rotate every 8 h, 24/7, and their pay and benefits is $12 per hour, then each shift costs IWC $240 per hour for preposting meetings or $720 each 24 h. If 1 h is spent in a preposting meeting per shift, that is costing the CSO's budget, thus IWC, $262,800 a year just to attend the preposting meetings!

As you can see, the numbers are substantial in both hours and dollars spent. When first looking at the process, one would never know that the costs would be so high. That is the advantage of using a process flow analysis coupled with an SMMP.

The next step of this process would be to brainstorm first with the manager and supervisors of the guard force and each of the supervisors with members of their guard force to look for alternatives to having these meetings and still be able to meet the goals of the meetings.

Not everyone affected by security metrics management will be supportive of it, even after they recognize the value of measuring performance and making changes.

This all seems all well and good on paper and in theory. However, would it be prudent for the CSO to anticipate that effecting a change to this process may not be readily accepted by the guard force? Do you actually think that most of the guard force will enthusiastically support a project designed to drive change and ultimately reduce costs? Resistance can be expected even when the effort is well-communicated and participation and feedback from supervision and the guard force is sought. Change is not an easy thing for many to face. It is particularly difficult when it may adversely affect the pay-checks of those being asked to make changes.

SUMMARY

The security guard force is often the most expensive component of the physical security function in any corporation, and therefore it should be a high priority for measuring processes through an SMMP.

As a CSO, you will probably find there are many ways to improve the performance, efficiency and effectiveness of your guard force. In this chapter, we used the example of paying security guards for briefings that could be considered a very inefficient process, as in this process the guards experience a substantial amount of downtime and are being paid for that downtime. This inefficiency can lead to additional costs incurred, as it can lead to overtime expenses or even additional hiring. Remember, labor costs in physical security are generally the biggest portion of the security budget.

SMMP techniques can successfully be used to assist in the analysis of security guard force processes. Data collections, flowcharting, and process analyses are tools to better help the CSO understand how things are working and where efficiencies (cost savings) may be gained.

As a CSO, one must be cautious when dealing with collective bargaining units (unions) representing your guard force. Assistance from the legal staff, human relations, and contract specialist is a must.

CHAPTER 10

Technical Security Systems

CONTENTS

INTRODUCTION

When we discuss technical security systems (TSS) we are referring to those systems that are used to support the physical security function. In particular, these systems are used to control and channel access to physical areas or other assets being protected. Generally, we are talking about intrusion detection systems (for our discussion we include fire alarm systems in this group; moreover, we will use the terms intrusion detection and alarm systems interchangeably in this chapter), access control systems, and surveillance systems (primarily video).

The use of intrusion detection systems, access control systems, and surveillance systems—the three primary security systems used and categorized as TSS—are constantly evolving, becoming more capable, sophisticated, and complex. With continued and rapid improvement of microprocessor technology, these systems are continuously improving.

The use of these systems, when effectively integrated into the entire physical security functional approach to assets protection, can provide reliable and cost-effective protection. One reason for this is that they can effectively supplement and in some cases replace the security guard force.

TSS FLOWCHARTS

As we have stressed throughout all of the basic security functions' chapters, once the drivers have been identified and the security functions defined and integrated into a security organization, the next step is to analyze them to fully understand how they work. The analysis processes begin with developing process flowcharts, starting with a macro process view and progressing down to the micro view, then working into the more detailed flowcharts. As discussed in earlier chapters, creating process flowcharts allows for each step

Security Metrics Management. DOI: http://dx.doi.org/10.1016/B978-0-12-804453-7.00010-0

in a process to be identified and analyzed. The development of process flowcharts (diagrams) facilitates a better understanding of how the process really works and where areas for improvement may exist.

As part of the TSS organization flowcharts, let us look at a more detailed flowchart of a TSS subprocess. Let us look at the subprocess of issuing the new International Widget Corporation (IWC) employee a badge for access to IWC's "Proprietary Research Area."

Can you identify areas where security metrics will assist in determining the efficiency of this process? What methods are needed to analyze the function to identify more effective and efficient processing methodologies?

TSS METRICS

Intrusion detection systems are one of the different and sometimes many layers used in the protection of a facility. How they are used and to what extent should be determined in the planning process. The site physical security survey should identify vulnerabilities, current and potential, and the layers of protection in use. When assessed against known or suspected threats, the need for intrusion detection systems to augment physical protections should become apparent. Typically, intrusion detection systems are used at the perimeter of a building, facility, or site. Moreover, alarms are used for highly sensitive internal areas requiring an extra degree or layer of protection.

Intrusion and fire detection systems generally save money by replacing people in many instances. In the long run, they are more cost-effective and efficient than just using guards alone. Before assuming additional risk, ensure that you consult with executive management and have them accept that additional level of risk. You can determine the cost–benefits and whether or not alarms will actually save you money by applying security metrics processes.

Alarm systems cost more to install than to maintain. The cost of alarm systems is greatest in the acquisition and installation phase. Once installed, maintenance and monitoring costs are generally much less than people cost. A return on investment can be calculated and used as a selling point on the value of alarm systems. Using alarm systems offsets the need for some guards. The savings in recurring guard costs can be compared to the cost for acquisition and installation of alarm systems. Remember, over several years, it is usually more cost-effective to use alarm systems to augment security than to rely on a larger guard force.

When seeking to use technical security measures to augment other physical security measures, it is important to develop a business case on the value of doing so. A key element in gaining budgetary approval for installing and maintaining a modern alarm system is through collecting relative data to demonstrate the cost–benefits (return on investment) and then develop an effective means of communicating that message. Essentially, tell the story using metrics graphically depicted. Some of the data collected and graphically depicted should include:

- Cost of current physical security—assets protection—processes (e.g., security guard force posts)
- Cost of a replacement security system of alarms
- Recurring costs of the alarm system, e.g., maintenance, responding to false alarms

Let us take the recurring costs factor and, using a security metrics management process, look at what those incidents and costs would be over the first year after installation and reported over quarterly periods.

Why would one want to track such incidents and costs? Well, there is the obvious reason, which is that the data is needed to validate assumptions and to conduct a cost–benefit analysis.

Collecting this type of data may also be useful in other areas such as dealing with the manufacturer or contractor if problems with maintenance and reliability occur. There is nothing like having real data to characterize a problem and help you state your case.

What supplemental measures should be used to enhance the Chief Security Officer (CSO) understanding of related problems and help state "the case"? Actually, there are several, to include:

- How long the alarms were off-line on each occasion—total time.
- How long was each alarm off-line on each occasion—individual alarm total times.
- The number of times a security guard was required to supplement/replace the cumulative downtime of the alarm systems—total costs in time and money. You may find that some of your security guard overtime is due to such problems.
- The number of times a security guard was required to supplement/replace each downtime of each alarm system—total in time and money.
- The time from alarm malfunction to call to maintenance provider and their response time. That is to say from problem identification to problem fix. This would be very useful as the contract should call for the maintenance contractor to respond within a specific period of time and maybe after the response window has been exceeded, the contractor incurs all additional costs, for example, security guard expenses as substitute for the alarm system.

Today's technologically advanced access control systems offer several options, or combination of options, for controlling access through automated systems. The key is to provide a reliable, cost-effective, integrated system. How does one go about determining how to do that? The first step is to evaluate your specific needs. Before a CSO pursues the acquisition and installation of a new system he/she must understand what is needed and what is available to meet those needs. The focus of the analysis, once needs (based on identified drivers) have been established, should be to seek the best value. That system that provides the best return on investment and passes the cost versus benefit test should be selected.

Establishing the criterion for system requirements will be easier if the CSO has a metric management program in place. If the important measures are already in place, the CSO should have the necessary data to help make informed decisions. For example, if the CSO knows the systems currently in use have a 3% failure rate, which is unacceptable, then the CSO must seek out a system that provides a less frequent failure rate (something <3%). If the CSO has no real idea what the current system's failure rate is, it would be difficult to definitively declare what an acceptable failure rate for the new system must be.

As stated above, the use of metrics to help analyze an existing process or system and develop future requirements can help the CSO attain the best system possible to meet IWC assets protection needs. Keep in mind, the end result will only be as good as the analyses and the quality of the security metrics used in the analysis process. In a flowchart the CSO should depict the badge issuance process; you will note that the process seems to be logical from the standpoint of an employee gaining access to a restricted area where the employee will work. What is missing from this flowchart and what should be included are the "What ifs":

- What if the person does not show up to get the access badge?
- What happens if the paperwork (employee's manager and area custodian's justification and approval for access) is not received and the employee shows up for badge processing?

It is of course a wasted trip, resulting in loss of productivity on the part of the new employee, which always translates into a loss of money. When you, as a CSO, look at the flowchart, can you find other areas where there may be opportunities for potential changes that may decrease process and productivity costs? If so, what are they? One place to start is with a macro view.

From the first step to the completion of the last step:

- How long does it take, on average, to process one badge request?
- Is this reasonable?
- Is this desired?
- Could, or should, the process cycle time be faster?

These are some of the questions a CSO should be asking and thinking about as part of the process analysis.

Let us assume that the IWC Technical Security Systems Manager is directed by the CSO to determine just how effectively or efficiently the process to make a new badge is and to be prepared to depict that process in terms of cycle times. Now, why would the CSO direct that action? In this case, someone has complained that his or her employee has not received their badge or did not receive it in a timely manner; thus, that employee's ability to perform their duties was adversely impacted. The CSO has heard this before and wants to get to the root of the problem and "fix it once and for all."

Once the cost in terms of time spent (which equates to productivity loss) is known, then one can further break down and analyze the process flow to determine potential areas for efficiency gains. Once discovered, a project plan could be developed to drive changes to the process. Moreover, security metrics should be developed to use in monitoring that process change to determine if it worked.

When conducting the analysis of the badge issuance process, a review of existing process documentation for some specified period of time (a year would be good) showing all the dates of requests, approvals, and issuance of the badge should be part of the analysis. Essentially, first the CSO must understand how the current process performs. All captured information (security metrics data) should be placed into an automated database making it available for better analysis. For example, the data that should be considered for entrance into the database, making it available for analysis, are as follows:

- Date of request
- Employee's organization
- Manager's name
- Date of employee manager's approval
- Area custodian's name
- Date the custodian approved the access
- Date the request was received in the Badge Office
- Date the badge was made
- Date the employee was notified to come pick up the badge
- Date the employee signed for the badge

This information would allow the "badge process improvement project team" to not only identify the times that were spent in each step of the process but also who the people were in the process and their organizations. Why is this important information? Every step in the process must be examined to ensure the process analysis is complete. Failing to do so may cause an important step to be missed, thus resulting in a missed opportunity to improve the process. It is possible the apparent problem may not be with the badge-making process itself but with a related or supporting processes, such as delays in management approval of the badge requests.

Once the project is complete—that is, the process has been changed—improved—and effective measures are in place—the CSO and his staff need to monitor performance watching for changes and trends. Improvements are expected; however, if they are not achieved, the CSO should revisit the process analysis phase. Furthermore, with effective measures in place, the CSO can use graphic depictions of the metrics to demonstrate success to all interested parties, for instance to executive management to demonstrate improvements, cost savings, or problem resolutions.

An important point to consider, as mentioned in an earlier paragraph, is the examination of related or supporting processes during the process analysis phase. The CSO may conclude after analyzing and measuring the badge-issuing process that the process itself works as designed and expected. The root cause of apparent process failures had more to do with supporting processes than the badge-issuing process itself. For example, if the badge-issuing process was actually performing better than expected but employees were failing to pick up their newly made badges in a timely fashion, then the delays or loss of productive time was actually caused by a supporting process. Process analysis will help the CSO uncover problems like this.

In the case of this project, the results of the study were depicted in graphic form and used as part of a presentation to management depicting the overall "State of Security."

When you look at the chart, what stands out to you? It should be the length of time that it takes on average to process an area badge request starting with the longest amount of time, second longest and so on. It appears that the main problem here is the length of time it takes the area custodian to approve access to the area under the custodian's responsibility, followed by the length of time for the employee to pick up the badge. Therefore, this is an area which can be targeted for process improvement.

When you analyze the chart, do you think anything is lacking on the chart or should be done differently? Generally speaking, no graphic depiction of metrics is perfect. In this case, it would be nice to have the actual numbers that are equated to the bars on the chart depicted on the chart under the bar chart itself. Some may consider it unnecessary or a "nice-to-have" item. The decision should be based on the:

- Point you are trying to make by briefing the charts
- Preference of the audience

What is the audience seeing when presented with graphic depictions? What is meaningful to them? A critical point to remember when building and giving presentations is to know your audience. Different audiences are interested in different aspects of a problem or performance. What executive management wants or needs to see may differ significantly from what the process owner needs to see. Show the right data to the right audience.

Before you present the briefing using the charts we continue to discuss throughout this book, present the briefing to your managers, assuming you are a CSO with staff, and it probably would be a good idea to also present the briefing to your entire department. That way, they know what is going on, what it costs to do their job and to elicit their support in finding more efficient ways to do the security business.

As the charts depict, the costs to make badges for one area at IWC, where a special access badge is needed, are not small. Improving how badges are made and thereby reducing cost will produce a savings for security and IWC.

> Efficiency, effectiveness, process improvement, and increased productivity are areas all security professionals should be concerned with. They are as much about good security as are the security processes themselves.

Surveillance Systems

Surveillance is an important tool for security in its effort to protect assets. Generally, surveillance is accomplished by using security guards or by using CCTV surveillance. Most frequently, a combination of both is used to achieve maximum observation and effectiveness for any facility.

As part of a site physical security survey, the need for surveillance should be identified. This need should then be assessed with the existing practice and capability. With this information, a plan for site or facility surveillance should be developed. The plan should consider the following:

- Purpose of surveillance: deterrence and/or observation
- Identify critical or high-risk areas
- Camera and guard mix
- Location of cameras
- Recording capability needed
- Need for hidden cameras
- Type of cameras needed: wide or narrow angle of view, low or high level of light; availability of solar-powered cameras should be considered.

Each has its strengths and each has its limitations. These limitations should be summarized in a chart that can be used as part of a security metrics management system.

Such a chart can be made part of a cost–benefit analysis and report when deciding whether or not to install some surveillance systems in lieu of security guard forces. It can also be used as part of an executive management briefing when discussing security budgets and what you as the CSO have done to reduce costs while still maintaining the minimum corporate assets protection program required.

What security metrics data collection areas would you want to implement for the surveillance systems? Generally, they would be the same as used for the alarm systems.

TSS CASE STUDY

A CSO was faced with significant budget cuts. The largest area of security expense was with the guard force. As the CSO, what methods would you employ to determine current costs and ways to reduce costs?

Although each CSO has their own method, the following process is one that has been known to work for the authors:

- Identify current security guard force costs, to include benefits.
- Identify the security driver for each guard post.
- Identify the processes used by the security guard force.
- Establish project plan(s) to look at security guard force efficiency possibilities in terms of hours and money saved.
- Analyze the possibilities of eliminating some security guard force posts.
- Conduct feasibility studies to determine what TSS have the potential for replacing what security guard posts.
- Conduct cost–benefit studies to determine if such security guard force "replacements" would be cost-effective in terms of start-up costs and recurring costs, e.g., upgrades and maintenance.
- Use a security metrics management program and develop security metrics charts along the way where one-time, short-term and long-term data collection and security metrics management processes would be beneficial.
- Make recommendations to management for a supplemental budget to implement projects for installing and maintaining TSS where they are shown to be cost-effective and also reduce costs of the security guard force.
- Implement and manage using the security metrics management system.

SUMMARY

TSS have become more reliable, sophisticated, cheaper, and useful in supplementing or replacing security guard forces. TSS's primary subsystems are the surveillance systems, access control systems, and intrusion detection systems (to include fire alarm systems).

Security metrics management process can help determine the cost–benefits of replacing security guards with TSS subsystems; as well as manage the cost–benefits of such systems on a recurring basis.

CHAPTER 11

Locks and Keys

CONTENTS

INTRODUCTION

Remember that a security department is an overhead function, albeit essential in supporting the mission, but nevertheless a security function that is supposed to provide services and support to the other departments. That means that the Chief Security Officer (CSO) should always strive to make the department as effective and efficient as possible. This of course means that the physical security subfunction of locks and keys must also be included in that philosophy. Locks and keys are the nonautomated part of the access control process, unlike electronic access control systems.

Locks and safes are common means used for protecting areas, electronic media, and documents. They can be used in different layers of the security profile or be a layer in themselves. Used inside a locked building or office, they are a layer of their own. Used on a perimeter gate or door they are part of that layer.

An axiom of locks is "the more secure the lock and the more reliable the lock, the more expensive it is." The type of lock used should be consistent with what the lock is being used to protect. For example, when protecting the contents of a desk drawer, a simple desk drawer lock should be used. When protecting sensitive trade secrets, a high-security locking device as part of a high-security safe or area ought to be used. Getting this equation right will ensure monies are properly spent. The locks, regardless of their type, whether key or combination should be purchased where volume discounts are available and periodic-purchase contracts can be put in place.

A risk management philosophy should be used, which includes a cost–benefit part, to set the criteria for what type of lock or safe should be used to protect what types of assets. In some cases, it may be more cost-beneficial to install an automated access control system, such as for a room or storage closet where the telecommunications systems are installed.

Security Metrics Management. DOI: http://dx.doi.org/10.1016/B978-0-12-804453-7.00011-2

LOCKS AND KEYS DRIVERS AND FLOWCHARTS

To maintain the integrity of this type of access control system, the issuance of locks, keys, combinations, safes, and security cabinets should be a centralized security function and not delegated to each International Widget Corporation (IWC) department. If so delegated, what happens when a person leaves and takes the lock or combination with them? How would the CSO know if a safe, security combination, lock or key met the risk management criteria for adequate assets protection?

In a centralized process, the lock and key function will, like many security functions, cause some productivity delays: container and lock request process, or picking up the combination or key for a container. For a lock combination, where the combination is always maintained by the security staff assigned to the lock and key function, no lock-picking expertise would be needed—a cost savings in time, training, equipment, and/or outsourcing.

With the locks and keys security functional drivers, like all the other security functions related to assets protection, there are certain drivers that apply to all and they are:

- Laws and regulations
- Risk management factors
- Executive management decisions
- Corporate policies and procedures
- Audit reports or security surveys showing deficiencies related to assets protection
- Contractual requirements which impact assets and their protection
- Input of employees relative to assets protection

These also drive the need for physical security and one way of meeting the need to comply with the security driver's requirements is the locks and keys subfunction.

The locks and keys manager or supervisor and CSO should be able to quantify some savings in time and thus productivity.

Once the locks and keys security drivers are identified and an evaluation conducted to determine the need for physical security to have a subfunction of locks and keys, then the next step would be to conduct an evaluation to determine if a separate organization is needed. In the case of IWC, a separate organization was deemed necessary. That being the case, the next step would be to document the locks and keys work processes beginning at a high level and then in subsequent lower levels.

LOCKS AND KEYS EXAMPLES OF SECURITY METRICS

Think about what data collection points you would want to have based on the overall locks and keys process flowchart. Then establish the data collection processes, to include data collection tools.

> The Locks and Keys security metrics management process would be based on the steps shown in its flowchart.

An ideal data collection system, if the CSO wants to know how long it takes to provide locks and keys service to employees, is to have a locks and keys office badge reader on both sides of the office door and have each employee do a "badge swipe" when they enter and when they exit the office. That way, the statistics could be efficiently collected assuming that all employees badge in and out of the office and not just hold the door open for other employees to enter or exit without swiping their badges through the badge readers. It also assumes that the system would be cost-effective.

One may wonder "Why sort by the hour for time in and time out?" The primary reason is to see when the busiest times of the days are. That way, the CSO can ensure that during those peak busy times, sufficient staff will be available. For example, perhaps most of the customers come in conjunction with their lunch time. The CSO should arrange for the security staff to take their lunch breaks during the nonpeak service hours.

Sorting by organization and personnel will show which organizations have the most requests and therefore the most needs for assets protection safes, containers, and similar equipment.

The items requested can be used as part of the locks and keys inventory systems (how many items, and of what type, are out) and then compare with those in stock. This helps with the ordering needs.

The ordering and delivery of these items should also be part of security metrics management process for locks and keys. This is especially true to document the delivery times so that one not only knows how far in advance to order new items but also to track contract compliance for delivery of items within certain time constraints.

LOCK AND KEY CASE STUDY

A CSO was hearing complaints from IWC managers that their requests for combination locks were not being met in a reasonable period of time. The

CSO discussed the matter with the locks and keys supervisor. The supervisor advised that the combination locks were ordered in a timely manner; however, their delivery was sporadic and sometimes there were long delays in deliveries.

The CSO was shown the spreadsheet tracking, their inventory levels and ordering levels process charts.

If you were the CSO, how would you handle this matter? Some would be quick to say that they would order more combination locks at one time and order them based on maintaining a higher inventory level. Yes, that would easily solve the visible problem, at least temporarily—maybe. However, it would not solve the systemic problem, which is the lack of on-time deliveries to IWC by the supplier.

Actually, it has been shown that, with a cost-efficient system, a minimum inventory is maintained and uses a "just-in-time" approach. In other words, do not tie up budget in inventory but rely on the items being delivered on time, when needed.

Using the security metrics approach and documenting the average request-to-delivery time periods, one can see if it is more or less than the IWC customer average. You as the CSO or your locks and keys supervisor can discuss the contract with the IWC contracting specialists and determine what obligations and penalties the supplier must meet as far as contract specifications detailing delivery times and penalties.

In this case, the delivery time specifications from request to delivery were vague, within 1 week of request receipt. Furthermore, the contract did not address any penalties for not meeting contract requirements. As the CSO, what would you do to remedy the situation or would you do the "quick and dirty" fix and then move on to another complaint and crisis?

The CSO decided to put in place a total process using the security metrics management system as the tool to correct deficiencies in the process and monitor the lock and key operation. The following information may be what should be collected:

- What item was requested?
- When?
- From what supplier?
- Cost?
- Delivery date?
- Average delivery time (from request to on-dock)?
- Average delivery time from on-dock to placement in the locks and keys inventory?

- How long in inventory before issuance?
- Non-timely deliveries brought to contracting staff's attention?
- Penalties incurred by supplier?
- If so, how much?
- If not, why not?

What data collection tools and charting formats would you use to track each of the above items? One type would be a timeline that would track individual and total average times from request, quantities to deliverance, and sign-out to internal customers.

Another supplemental way to handle this is to contract to multiple suppliers. It is of course assumed that the bids for this work were and will be based on the lowest bidders who meet contractual specifications, in other words, the suppliers must be both responsive and responsible.

By using a security metrics management approach, not only will the process eventually be made more efficient, but the customers will hopefully be satisfied with the process time for their requests for services (lock or key) and IWC will in turn gain in productivity-efficiency of its employees since they will not be "wasting time" working on lengthy requests for services from the locks and keys function, getting their necessary security containers, locks, and other equipment in a timely fashion. Also, suppliers who begin to incur penalties will be faced with the need.

SUMMARY

The locks and keys security function is a subfunction of the physical security organization and is a "key" (no pun intended—well, maybe) function to support the protection of corporate assets. The function is also time intensive and detracts from employees' productivity since they must spend valuable time processing a request for services and visit the locks and keys office to obtain their locks, keys, and other related equipment.

Using a security metrics management approach, as with other security functions, the CSO can analyze the locks and keys function for areas to improve on. Basic cycle time and process data can be collected, analyzed, and the changes made to the process to improve efficiency and effectiveness, thus contributing to the productivity of the security function and the productivity of the security department.

CHAPTER 12

Fire Protection

CONTENTS

INTRODUCTION

Many security organizations are also responsible for the fire protection program. Responsibilities for security and fire protection, when coupled, are generally a good fit as both are concerned with corporate assets protection. Although the disciplines of security and fire differ, there are significant points of intersection. Both organizations serve in a preventive and protection role and both are involved in the business of emergency response and assets protection—especially the protection of people and property. Security, particularly the security guards' functions, fire prevention and suppression (FP&S) and to some extent, the safety function, all see their missions intersect in the areas of assets protection and emergency response. Furthermore, this intersection also continues into the area of contingency planning and employee asset protection awareness programs.

Under ideal conditions, FP&S are best handled by a professional fire department. However, having a company fire department is often not an option. It costs money to maintain a fire department staffed with professional fire fighters. Small- and medium-size organizations usually do not have large enough statements of work to warrant having a full-time fire department. Under these conditions, many companies rely on others to perform the duties of FP&S.

Fire protection programs usually separate into two areas: fire prevention and fire protection or suppression. Fire prevention refers to the effort to control the environment in order to reduce the possibility of fires from occurring. This includes good housekeeping practices, awareness training, and fire safety inspections. Fire protection or fire suppression refers to the methods and equipment used to extinguish and suppress a fire.

Throughout this discussion, the terms *fire protection* and *fire suppression* will be used interchangeably. At International Widget Corporation (IWC), the

Security Metrics Management. DOI: http://dx.doi.org/10.1016/B978-0-12-804453-7.00012-4

new Chief Security Officer (CSO) had inherited the IWC FP&S program. Since this is the first time the CSO had this responsibility, the CSO researched the topic and then implemented a security metrics management program (SMMP) to provide an overview and management oversight process for the FP&S program.

FIRE PROTECTION DRIVERS AND FLOWCHARTS

The US Occupational Safety and Health Administration (OSHA) requires employers to implement basic measures such as providing fire-fighting equipment and employee training in order to help prevent injuries and death from a fire. This requirement is one of the primary assets protection or security drivers for this function.

The IWC FP&S processes had never before been analyzed to assess their effectiveness. To begin that process, the CSO planned to review all process flowcharts. Much to the CSO's dismay, none of the FP&S processes had ever been depicted in a process flow diagram or flowchart. Therefore, the CSO tasked the fire department chief to develop process flowcharts for all of his FP&S processes, starting with the macro view of the processes of prevention and suppression and work down into micro views of all subprocesses. As part of this task, the CSO asked that all drivers for these processes be identified and documented. For operations within the United States, the primary driver was OSHA. International facilities identified their drivers to be the need to protect assets for the corporate owners, local laws and regulations, and the corporate policies and procedures for IWC.

The US FP&S manager began with a high-level flowchart and subsequently added more detailed flowcharts.

After completing these process flowcharts, the fire chief must then continue to refine them in more detail. For example, under "Preparation," the chief must provide further subprocess detail depicting the steps involved in preparing for an inspection, such as review last report, identify findings, identify inspected organization's response to findings, and prepare notes to validate inspected organization's reply to findings.

Under "Report," the detailed process of preparing the report, dispatching it to the inspected organization, establishing a response due date, and arranging follow-up inspections as necessary would be addressed, as well as of course input the data into the SMMP.

Also under "Report," the IWC CSO directed that the FP&S chief initiate a project to analyze the reporting process. This includes report preparation, presentation, and delivery. How much time and resources are being used to prepare, present, and distribute the inspection reports, and should report preparation and delivery be a web-based process, thus affording all participants to ability to contribute online in near real time?

Should the inspection process itself be changed to a random and unannounced process? This unknown factor should keep organizations prepared, as they will never really know when they can expect to be inspected. Another possibility is to rely on self-inspections with limited follow-up by FP&S personnel to ensure the self-inspections are conducted in accordance with IWC guidelines and maintain an appropriate level of integrity—or possibly convince the audit specialists to add this item to their auditing checklist when evaluating IWC departments' compliance with IWC policies and procedures. This approach would be a cost-effective way for the CSO to get the job done.

This is not to say that this is necessarily a good idea but simply one offering potential for possible savings—or is it? Remember that cost in terms of time and lost productivity, when transferred from one organization to another within IWC, is not really a savings. This is something to keep in mind when trying to save resources for IWC while meeting all security service and support requirements. Also, outsourcing this function should be considered.

FP&S METRICS EXAMPLES

If fire safety inspections are to be conducted regularly, their frequency will depend upon many factors to include the conditions and type of operating environment. At a minimum, inspections should have as their objective the following:

- Identify fire hazards, conditions, and housekeeping discrepancies.
- Determine compliance with applicable laws and government agency requirements.
- Identify contract provisions and insurance carrier requirements.
- Ensure corporate and business area loss prevention programs are carried out as directed.

The inspections should be conducted in the context of the following parameters:

- Establish and maintain inspection methods and intervals, consistent with corporate interpretations of federal, state, and local government codes, laws, and standards applicable to FP&S matters.
- Inspect work areas periodically to ensure compliance with fire regulations. Hazardous areas may require more frequent inspections as

determined by fire professionals and regulation. Also, the following specific FP&S areas should be inspected:

- Fire pumps
- Fire water supply
- Main drain: residual and static flow tests
- Sprinkler systems
- Fire extinguishers and fire extinguisher records
- Fire doors
- Fire walls
- Complete documentation (records) of fixed fire suppression equipment, including test and maintenance records, and fire insurance inspection reports need to be kept. They are important for the process of fire prevention planning and, in the event of a fire, will be subject to review.

The security metrics would be used not only to collect and track inspection data and results (satisfactory, marginal, and unsatisfactory) but also by types of deficiencies. They could be correlated and if analyses indicated some areas where the majority of deficiencies were the same, such as outdated fire extinguishers, then a project could be implemented to look at the systemic problem which allowed this to occur.

In addition, other deficiencies within multiple IWC departments could be addressed through the security education and awareness program—emphasized in briefings, through awareness material, broadcast announcement via the IWC internal network, or other security awareness methods.

So, inspections are a candidate for data collection and process analyses as part of the SMMP. For example, overall inspection results could be tracked, as well as other data that would be useful in managing the FP&S program and depicted through the use of metrics (graphically depicted) to demonstrate to executive management and outside compliance agencies (i.e., OSHA) the effectiveness of the program. Even in areas where the program may be lacking or be ineffective, if an understanding of the problem along with a plan to take corrective action is shown to a compliance agency, chances are any penalty or fine would be mitigated.

Having a comprehensive SMMP in place will also be useful for IWC when dealing with insurance companies. A proactive program may help reduce insurance costs and thus show direct benefits. It also would be very useful in the event of a fire when executive management and the insurance provider demand to know the "who, how, where, when, and what" of a fire, instead of you as the CSO saying, "It's not my fault! We did all we could to prevent the incident!" You can show a pattern of aggressive inspections, corrective actions, and so forth.

Fire alarm and suppression systems are a critical component of a fire prevention and protection system. Fire alarm systems detect fire and alert employees. They are similar to intrusion detection systems in that they, in and of themselves, do not prevent, but they only alert. Unless the alarm is false, when the fire alarm system alerts, it is due to fire or the elements of fire (smoke or heat). Like all equipment, automatic fire suppression systems require periodic maintenance to maintain serviceability and effectiveness. Maintenance must only be performed by qualified and certified personnel. When automatic fire systems are taken out of service, usually for maintenance, a fire watch should be instituted, using trusted employees, in order to be able to react in the event of fire.

False fire alarms, fire alarms' periodic and troubleshooting maintenance, and related costs are examples of fire protection-related topics where a CSO should apply security metrics management techniques.

By collecting data on problems, the CSO can use the information to better help understand overall performance and identify problems. Moreover, with data collected over time, trends will become apparent (both positive and negative), which the CSO can use as a tool to help drive process change and even changes to the specifications of the contracts with service providers. For example, if the alarm service contract specifies specific times for alarm response, maintenance call response, details on warranties, and costs, to include costs for repairing the same problem over a short and designated period of time—1 month, for example—the CSO can use performance security metrics to drive the service provider to greater efficiencies and better performance.

False alarms are expensive. All false alarms require a response as they cannot be considered false alarms until an authorized responder makes that determination. This is a case where the life of employees or other building occupants are at risk. There is no margin for error here.

False alarms occur for a variety of reasons, including faulty equipment, environmental elements, and employee mistakes (accidentally activating an alarm). One way to help mitigate the negative impact of false alarms is to have in place a fire warden (employee designated for a specific area) program where trained employees respond first to determine if the alarm is false.

The area that the fire warden covers, however, must be a small area so that the warden can respond to the area immediately. This again is a costly element of assets protection and this process should be incorporated into any SMMP for this security function. However, as in other instances, this too costs money. Do the benefits outweigh the costs? Use an SMMP coupled with its risk management aspects to find out.

Other data collection and charting processes should of course be set up for the other subfunctions of the FP&S organization. These would include:

- *Fire Prevention Education and Training*: The metrics processes used can be the same or at least similar to the security metrics management system used for the entire security education and awareness training program.
- *Department Evacuation Plans*: Plans must be tested. Security metrics can provide a database as to the number of tests conducted, the departments involved, number of personnel, costs, results, and any follow-up action.

Again, the first step is to identify the drivers, identify the organization structure based on what will work in your environment based on those drivers, then develop detailed process flowcharts of macro and micro processes, determine what data should be collected for meeting what objectives, and then develop the data collection input documents, analysis processes, and finally then the charting and the management oversight and briefings can occur.

CASE STUDY—OUTSOURCING FIRE PREVENTION AND/OR SUPPRESSION

The fire prevention and protection process is an outstanding candidate for outsourcing. Since IWC does not have as part of its core competencies professional fire prevention and protection capabilities, the CSO will be considering it for outsourcing. The CSO knows that the fire prevention program must be established and, to reduce the risk of fire, it must be effectively and efficiently established. However, it may be more cost-effective for someone else to do it.

Protecting IWC from damage and loss from fire cannot be ignored. However, the CSO reasoned that it can be done cost-effectively. Maintaining a capable staff to perform fire prevention and protection duties is costly. Skilled fire-protection professionals require regular training and certification. Finding capable outsource providers of fire protection services is not difficult. There are many capable companies, local and national, with resources and experience necessary to perform fire prevention services—which of course, include the local city and county fire departments.

As the CSO wanting to determine if this function should be outsourced, what metrics would you use and how would you go about determining if outsourcing was a viable option?

SUMMARY

FP&S programs are an important part of IWC's ability to conduct business. No one is immune to the threat or hazard of fire. Often the responsibility for administering fire prevention programs falls to the security department.

Although the discipline of security and FP&S differ, there are a few common denominators. Both are in the business of assets protection and emergency response. Therefore, coupling security with the FP&S process can be very effective in the overall assets protection program. It stands to reason then that the SMMP can also be useful in managing an FP&S program.

CHAPTER 13

Event Security

CONTENTS

INTRODUCTION

Conventions, trade shows, shareholders' meetings, ground-breaking ceremonies, new product introduction events, and employee recognition activities are just some of the regular special events that International Widget Corporation (IWC) holds. Some of these events are held on IWC property, in IWC facilities and buildings. Many are not. Some are held in high-profile locations, such as trade shows held in major convention centers located in large cities, while others are held in international venues. Some of these locations are easy to secure while others are not. Generally these events involve large numbers of people converging on a single site. Often local and sometimes national media will be involved.

Special events are held for many reasons and in many locations. Regardless of where and when they occur, and even though the host facility, such as a convention center, may provide overall security, some form of security support for the corporation's specific location may be necessary. From access control to executive protection, security for special events can vary widely. Whatever the level of support required, the chief security officer (CSO) or event security coordinator (ESC) will take the same approach for assessing the security needs and requirements for each event. For the IWC, the CSO will be serving as the ESC.

EVENT SECURITY DRIVERS AND FLOWCHARTS

The event security driver is, as usual, the need to protect corporate assets. The specific assets protection techniques applied to events should be based on a risk management approach—threats, vulnerabilities, and risks considerations applied to the location of the event—and also the public relations considerations, such as stockholders in attendance going through metal detectors, being searched, etc.

Security Metrics Management. DOI: http://dx.doi.org/10.1016/B978-0-12-804453-7.00013-6

Event security flowcharts are prepared like any other flowcharts. And as is the case with the Executive Protection flowcharts, these should only be distributed to those on the event security team and based on the "need-to-know" principle.

EVENT SECURITY METRICS

The objective of event security is obviously to protect the event—i.e., the people attending the event, facilities, other assets, and any sensitive information not in the public domain. Much like executive protection, event security has obvious measures of success:

- Did something occur at the event to adversely affect that event?
- Did something go wrong?
- Was the event interrupted?
- Was harm caused to corporate property or personnel?

If the answer to all of these questions is no, generally the event can be considered a success.

It is possible for something to go wrong without notice. For example, the disclosure of sensitive information may not be discovered by the IWC CSO until a later date, if at all. A competitor (or adversary) could gain access to sensitive information, thus providing that competitor an advantage. That advantage may never be directly related to disclosure of information at the special event. (This situation is not the most likely, as most special events are not places where sensitive information is usually held.)

Like the Executive Protection function, event security metrics are generally limited to those related to successful planning and implementation of Operational Security (OPSEC), the security plan, budget versus actual costs.

As for applying security metrics management techniques related to OPSEC, one may have a difficult time since OPSEC is as much a process of planning as it is a process of action. Planning is a process that can be measured, but it is best measured in the context of comparison to prior plans and their success and failure:

- Were all relevant possibilities considered?
- Were all possible scenarios addressed?

The best measure of a new plan is to compare it with an earlier successful plan.

As for the event security budget, that is somewhat easier, as one can easily use a spreadsheet to:

- List budgeted line items
- Collect the actual costs for those line items

- Identify the variances
- Determine the reasons for the variances
- Readjust the event security budgetary process for future events

Note that in your actual spreadsheet, every process step should be identified in an individual spreadsheet cell and its costs tracked. This can be done by using the process steps noted in a detailed event security flowchart.

EVENT SECURITY CASE STUDY

Since the IWC is about to hold its annual shareholders meeting, the IWC CSO has begun preparing a security plan for the event. As a CSO responsible for such an event, how would you go about preparing a security plan and its budget for such an event?

The IWC CSO did the following, based on the detailed flowcharts and other security metrics management program related documents:

- Reviewed the security plan and budget that was used for last year's shareholder meeting.
- Reviewed the following for applicability to this year's event and begin specific planning on a budget for each area:
  - Advance work—preevent planning and analysis activities
  - Establishing a security operations center
  - Establishing physical security controls
  - Establishing information security controls
  - Identifying the need for personnel and executive protection
  - Preparing for likely contingencies.
- Reviewed the variances between last year's event budget and the prior year's event budget (NOTE: determine why the variances occurred and take action accordingly). Cost overruns should be analyzed and, if needed, applied to the next event security budget. If due to poor planning, analyze the process and all costs involved.
- Developed a new budget based on the historical budget information, taking into consideration the variance in the event for this year and prior years, postevent security analyses, etc.

SUMMARY

Providing security for a special event or offsite activity is part of the CSO's responsibility. These events are usually short in duration and occur with some degree of frequency and are important to executive management, stakeholders, and IWC.

Working to ensure the event proceeds as planned without a security incident or an event disruption is essential. A single incident may damage or ruin the entire event, as well as the corporation's image and public relations. The IWC CSO must develop a security plan for all events that includes a separate budget.

Security metrics management techniques do not play a major role in such events, except that the success of an event means no adverse incidents took place. You do not need a security metrics management system to tell you that.

The primary areas where security metrics management plays a role are in the planning development and budgetary processes. Comparing new plans against old does provide a level of assurance all relevant issues are addressed—assuming it was done correctly the last time.

Measuring budget against actual costs not only helps you determine how well you planned for this event, but it also provides an overall evaluation of how you plan security budgets in general. The lessons learned from budget planning for events should also be incorporated into other budgetary planning processes.

CHAPTER 14

Executive Protection

CONTENTS

INTRODUCTION

Executive protection is the application of protective measures to reduce the risk to executives and mitigate threats. It is a proactive effort, not a reactive effort. Although there are reactive elements, proactivity is the only way of effectively providing protection. Proactivity calls for advance preparation, so in situations where reaction is necessary, it will be automatic. Part of establishing an executive protection program is to identify the executives requiring protection. How you do this depends on many factors, including the following:

- Culture of your corporation
- Threat to your corporation
- Negative profile of corporation (e.g., some corporations have a negative profile attracting the attention of persons in opposition to their mission, an example being oil corporations as they are challenged by radical environmental organizations.)
- Habits of the chief executive officer (CEO) and other senior executives (e.g., travel to high-risk areas)

At International Widget Corporation (IWC), there are different levels of executives. Some are corporate executives and some are business unit executives. They hold different titles. Some hold the title of president, some vice president, while others are directors. Yet all are not deemed critical to the corporation. This is not to say they are not important to IWC. It is to say that only a critical few are so important as to warrant additional protection beyond what is afforded to other employees.

For executive protection to be effective, the protector (which may be an individual or a team) must know the following:

- Who is being protected?
- What are they being protected from?

Security Metrics Management. DOI: http://dx.doi.org/10.1016/B978-0-12-804453-7.00014-8

- When are they to be protected?
- Where must they be protected?
- Why are they being protected?
- How must they be protected?

The executives at IWC who have been determined to require additional protection are the following: CEO, chief operating officer (COO), chief information officer (CIO), and chief financial officer (CFO) and the heads of each separate business unit. All other executives are protected in the same manner as any IWC employee is protected. How this determination was made focused totally on the criticality of each executive to the performance of their separate business units and to the success of the corporation. In other words, what would be the impact to the corporation if the executive was suddenly not available—permanently? Loss of expertise, continuity of leadership, and customer goodwill must be considered in the assessment. Moreover, the level of public exposure and international travel of each executive must be assessed to determine any operational risk.

In today's global business environment, there are many types of threats faced by corporations. All threats require some degree of attention and mitigation. The potential threat to business executives is no less of a threat to the total corporation than any other business threat and, in some circumstances, may be even more cause for concern.

High-profile executives leading corporations in controversial industries may find themselves potential targets since they are, or at least they represent, the leadership of that corporation. With some high-profile executives, they essentially become a personification of the corporation. This places them in a higher risk category as their visibility significantly increases.

The purpose of an executive protection program is to reduce the likelihood of an attack occurring against corporate executives, thus reducing the overall risk to IWC's business. This is a critical factor. Protecting high-profile or high-risk executives also protects the corporation. Furthermore, eliminating the potential for harassment or embarrassment to the executive is essential. Although embarrassment or harassment does not cause physical harm, these situations can impact the effectiveness of the executive, thus having a negative impact on the corporation. Eliminating risk is neither possible nor practical, but mitigating it is both.

An executive protection program should be designed to facilitate executives living and working safely and moving about efficiently. They cannot be locked up. Their ability to move around and work freely is essential to their performance. Facilitating this mobility in a safe and secure manner is the objective of an effective executive protection program. In essence, the protected executive is being made into a hard target.

EXECUTIVE PROTECTION DRIVERS AND FLOWCHARTS

As usual, in developing a security metrics management program for this security function, a high-level flowchart is first developed and subsequently more detailed charts are developed, derived from their key functional process steps. By now, you should have first been able to identify the drivers for this function before proceeding to the flowcharting. They are of course based on the drivers requiring the protection of assets and in this case the assets are people.

It is very important that all process documentation and related information (flowcharts, metrics evaluations) be kept within the executive protection team (even from other security specialists who are not part of the executive team or team support functions) so as to prevent unintended release of operational procedures and plans. This is especially important in today's high-risk global environment.

Terrorism, kidnapping, hostage situations, harassment, or embarrassing situations are a greater threat to the corporate executive, particularly Western executives, than ever before. Therefore, discretion, privacy, and good operational security techniques are essential for success.

If the processes used to protect executives were made public, those miscreants wanting to harm or exploit a corporate executive could develop an advantage to their misdirected exploits.

Detailing the protection processes in flowchart form will better assist the executive protection specialists in preparing and understanding each process, as well as the time it takes on average and the associated costs of each step in the process. Moreover, the process diagrams are useful when establishing checklists for use by the executive protection specialists and the executive (and family where appropriate) being protected. It is also a method of looking for weaknesses in the process which could lead to previously unidentified risks to the protected executive.

EXECUTIVE PROTECTION EXAMPLES OF METRICS

As the IWC chief security officer (CSO), you must be intimately involved in executive protection even though it may be outsourced. When the executives are being protected, there must be constant communication and feedback between the CSO's office and those providing the protection. This would include pre-protection meetings to review the itinerary of the executive and coordinate protective measures. Furthermore, the protective measures must include contingency or emergency procedures. Below is a checklist of things, which were developed from a detailed flowchart of the function, to consider and address as part of an advance assessment:

Getting there:

- Where is the venue?
- How will the executive get there (mode of travel and route)?
- What will they be doing while they are there?
- How many locations will they visit?
- Who will they be with?

Once they are there:

- Is anyone in the area considered a threat?
- Is there a threat to similar people (other business executives)?
- What type of information about the protectee is available and accessible?
- What is going on at the location, today and tomorrow?
- Who will be there? Any other high-profile persons?
- What type of assistance is available?

Conduct a site survey to identify issues, routes, and layout of the following:

- Meeting location
- Hotels, resorts, and restaurants
- Private residences
- Convention center
- Emergency facilities
- Airports

Identify all concerns and issues:

- Is there anything out of the ordinary to be concerned with?

Departure:

- What is the return route (direct and fast)?
- Establish alternate routes

Emergency services:

- Where are the nearest medical facilities?
- Will medical personnel be available on-site where the protectee will be visiting?

Once the advance effort is complete, review findings and share them only with those people who have a need-to-know. Also, be sure to brief the executive or executives involved so they are familiar with all plans and issues.

Once that checklist is established, details are added for each trip. After each trip, the time and costs needed should be collected and entered into a database or spreadsheet. This will be very useful when planning the time and costs of future executive protection budgets, trips, and the like.

Remember the checklist also has value beyond ensuring nothing is overlooked or forgotten. It can be used to help establish discrete and total cost of the protective measures or the operation. A cost can be associated with every protective practice. Cost and effectiveness can be evaluated post-event, allowing the protection specialists or the CSO to determine which protection measures are the most effective.

In the case of this security function, the CSO can consider additional security metrics and associated charts for tracking costs for this security function. The real determination of the success or failure of this security function is whether or not the executive being protected has been harmed, harassed, or his/her work disrupted. One does not need a chart or data collection for this; however, as with other functions, security metrics data collection on time and costs are always valuable, as it would not be unusual for an executive to want to know what all this protection is costing the corporation.

EXECUTIVE PROTECTION CASE STUDY

The IWC CSO was advised that one of the company executives identified as requiring protection by IWC executive management and the CSO has refused that protection, except when traveling internationally, and only to countries considered hostile to Western interests. As the CSO, what would you do?

The IWC CSO scheduled a briefing for all company executives identified as requiring some level of protection. During that briefing, the rationale for executive protection was briefed to include the threats, vulnerabilities, and risks associated with executives in general and IWC executives specifically. All were briefed on the overall executive protection plan and how protective operations would be conducted. It was expected that all would agree to accept the protection as planned.

The CSO also anticipated that the executive refusing protection support would be swayed to accept that protection by his/her colleagues. If not, the CSO would prepare an "at-risk" memorandum to be signed by the executive refusing protection, sending a copy to the CEO. The CEO has the prerogative to waive the protection requirement or direct that the protection be provided as a benefit to IWC.

The same methodology would be used for those who wanted more protection, such as protection beyond what is normally provided to a senior executive of a global corporation. The perception of need for executive protection will vary from executive to executive. Most executives will readily accept (or expect) a reasonable level of protection appropriate to the threat or situation. In some cases, a few executives will expect much more, generally based on satisfying their ego as to their "importance" more so than knowledge of real and actual threats. The CSO must find a way to work with those executives and convince them to change their expectations. Ultimately, the CEO and/or board of directors make the final decision as to who is to be protected.

SUMMARY

Security metrics plays a limited role in supporting executive protection operations; nevertheless, it does have applicability and value. Executive protection is primarily a physical security function and therefore often led by the manager or supervisor of the physical security organization. Simplistically speaking, if the executive being protected is not harmed, embarrassed, or harassed, then the operation was a success. If an adverse event did occur, then it was a failure.

Providing protection for key executives is both a good and essential business practice. Maintaining the leadership continuity of IWC ensures that IWC will be able to continue uninterrupted business operations. How protection is accomplished, who is protected, and to what degree they are protected must be based on a thorough threat and risk analysis conducted by the CSO or other qualified persons such as an executive protection specialists or consultant. The final decision as to who is protected is up to the CEO and/or the board of directors and not the CSO or the person to be protected.

Balancing threats, risks, vulnerabilities, and costs will be necessary to effectively develop and implement any executive protection program. Security metrics can help track costs and analyses may find ways to provide for more effective and efficient protection; however, in this case, one must lean toward effective protection over efficient protection since people's lives are at stake.

CHAPTER 15

Investigations and Noncompliance Inquiries

CONTENTS

INTRODUCTION

Investigations and Noncompliance Inquiries (NCIs) are security functions that, at some corporations, may be candidates for outsourcing. At this time, both are internal functions at International Widget Corporation (IWC). At IWC, these two functions are much alike. However, the primary difference is scope and magnitude. That is to say, an NCI is used to describe an investigation which is conducted due to a violation of corporate policy or procedures where there has not been a violation of law. At IWC an investigation is generally a much more complex process associated with a more serious situation and may involve a violation of government law or regulation external to the corporation.

One primary reason for the differentiation is for public relations purposes. When one hears that an investigation is being conducted, it sounds more serious than if one hears an NCI is being conducted. In addition, an NCI may be used as a preliminary inquiry to assess if something is wrong and requires further investigation. An NCI may be conducted by almost any security professional or member of management. An investigation requires someone skilled in the techniques and processes of investigations and with a working relationship with different governmental investigative organizations (local, state, and federal).

INVESTIGATIONS AND NCI DRIVERS AND FLOWCHARTS

Of course, after the organizational structure of the investigations and NCI function is identified and charted, its primary drivers are identified and graphically depicted. In this case, the complaints and allegations from various sources are considered the security drivers.

Security Metrics Management. DOI: http://dx.doi.org/10.1016/B978-0-12-804453-7.00015-X

Remember, one important driver for the investigations and NCI process is the total number of IWC employees: The more employees, statistically the larger number of employees who will violate IWC rules or government laws. Another is the number of requests for support from other organizations (such as the legal and ethics staffs and their need to have investigators support their processes). Remember that the investigations and NCI organization is also a service and support organization and as such must provide professional support to other IWC organizations when that service and support is requested and determined to be warranted.

INVESTIGATIONS AND NCI EXAMPLES OF METRICS

There are many different security metrics that a chief security officer (CSO) can use to help understand, assess, and manage the investigations and NCI processes. A problem that may face the CSO, as with all other security metrics, is determining the most useful metrics. When in doubt as to the most valuable metrics, the CSO can start by identifying as many as possible and then sorting through them to determine which offer the most utility.

As previously stated in this book, the CSO may first want to develop a process flow diagram depicting the macro process and then develop flow diagrams for the investigative subprocesses or micro processes. Once accomplished, the CSO can begin to develop points for different processes measures. An example of such a data collection list for investigations may look like the following (one for the NCI function would be almost identical):

- Number of investigations opened per month
- Number of investigations closed per month
- Number of investigations pending per month
- Average time used to conduct an investigation
- Average cost in terms of investigator's time, IWC employees' time, administrative time, and cost of resources used
- Same information as above broken down by type of investigation
- Same information as above broken down by quarters, year, and multiple years
- Identification of the IWC departments where the incident took place
- Identification of the IWC departments where the subject (employee) of the investigation was assigned
- Number of allegations proven correct
- Number of allegations proven wrong
- Subject employees of investigations position and job code

- Type of investigations broken down by departments
- Department information broken down monthly, quarterly, annually, and multiple years
- Associate a cost chart with each of the above charts, where applicable

By using this approach, one can begin to get a sense of the type of information that offers potential for developing useful metrics. Furthermore, the CSO can relate the potential data points to what he/she needs to know. For example, if the CSO is attempting to determine the average time to conduct an investigation, tracking the time taken to complete all steps from the opening of an investigation to the closing of an investigation will provide that data. The CSO can further analyze that information by sorting investigations by type. An investigation into the theft of a physical asset, on average, may require less time than an investigation into misuse of information systems.

Metrics developed and used in the investigations and NCI process may provide value beyond the investigative processes itself. Trend data may be developed and used to drive changes in other routine security policies, procedures, and processes. For example, if investigative trend data reveals thefts to be occurring during a specific time frame, additional protective measures could be implemented during that vulnerable period to either prevent the thefts or catch those committing the thefts.

The information gathered can be used proactively to reduce the number of incidents requiring investigations, thus reducing the overall workload for security investigators. Learning from security incidents helps prevent their occurrence in the future.

In Appendix are a few examples of graphically depicted security metrics charts that a CSO may find useful in the effort to assess effectiveness of the investigations and NCI process and better manage the organization.

Remember that as the employee population increases, generally so does the investigations and NCI. This is an important chart in that it depicts a trend. From this graphic depiction, the CSO can conclude a continuous yearly increase in the number of investigations. This trend should drive the CSO to analyze the investigative data to determine why. Action taken, or not taken, will be influenced by the results of the analysis.

Process measurements can tell much about a process. The type of measures used should correlate to what the CSO wants or needs to track and understand. The ultimate goal for the CSO should be to understand what is occurring that drives the need for investigations. Can those drivers be changed to such a degree as to eliminate or reduce the need for investigations? Measuring will also tell the CSO if the changes made had any effect on the process. Of course, cost issues must always be considered.

INVESTIGATIONS AND NCI CASE STUDY

As a CSO, you decided that it would be a good idea to use the security driver's metrics used for tracking the number of employees, the number of inquiries, and the number of investigations conducted over time. You have gone through the analytical process to make that decision based on answering the how, what, why, when, who, and where questions noted below.

- Why should this data be collected? To determine the ratio of employees to the workload; thus manpower requirements could be forecasted over time.
- What specific data will be collected?
  - Total number of IWC employees
  - Total number of NCIs
  - Total number of investigations
- How will these data be collected?
- Total employees: The collection will be accomplished by taking the total number of paid employees from the human resources department's master personnel database file.
  - Total number of NCIs: This information will be gathered by the unit coordinator from the unit's NCI database file.
  - Total number of investigations: The unit coordinator will also gather this information from the unit's investigations database file.
- When will these data be collected? The data from each of the previous months will be compiled on the first business day of each of the following months and incorporated into the crime investigations drivers' graph, maintained on the investigations and NCI's administrative information system.
- Who will collect these data? The data will be collected, input, and maintained by the unit's coordinator.
- Where (at what point in the function's process) will these data be collected? The collection of data will be based on the information available and on file in the investigations and NCI's database at close of business on the last business day of the month.

The CSO and organizational manager will analyze the NCI data, for example, to determine:

- The reason for each employee's noncompliance
- The position and organization of the employee
- Their seniority date
- Identification of the patterns
- Main offenses

That information would then be provided to the project team assigned to the goal of decreasing the need for NCIs and investigations. Based on that information, the briefings would be updated and more emphasis placed on those areas that caused the majority of problems.

Remember, there are numerous types of graphic depictions of data that can be a great tool for management. They include bar charts, pie charts, and line charts. The charts can be monthly, quarterly, weekly, or annually. The timeliness of the charts should be dependent on the manager's need for the information.

The key to the data collection and their related graphic depictions is to look more at trends than monthly numbers. The goal is to continue to maintain and improve on positive trends. Negative trends should be analyzed for systemic causes and project plans implemented to reverse the negative trends. The metrics could then be used to monitor the process and to determine if process changes actually cause the reversal of the negative trends. If not, then new analyses and rethinking of the problem are called for.

The organizational manager, in coordination with the CSO, began this process by of course identifying the drivers requiring the functions to be performed. Then the processes were flowcharted and a process analysis summary was developed to help provide a high-level view of the process.

That process summary included the following information:

- Security department: Investigations and NCIs
- Process definition: Provide professional investigative services in support of IWC and its customers
- Subprocesses:
  - Conduct investigations
  - Conduct NCIs
  - Conduct crime prevention surveys
  - Conduct crime prevention special briefings
- Requirements and directives that govern the process:
  - Corporate assets protection program
  - Contractual security requirements
  - Position descriptions
  - Corporate policies
- Suppliers:
  - IWC employees
  - Customers
- IWC management
- IWC customers
- Input:
  - Complaints
  - Allegations

    - Requests for assistance
    - Security requirements
- Output:
    - Investigative reports
    - NCI reports
    - Inspection reports
    - Security assessment reports
    - Briefings
    - Testimony
- Key metrics:
    - Subprocess 1: Case totals year-to-date and 5-year trends, case aging charts
    - Subprocess 2: Crime prevention surveys completed, results, and cost–benefits charts
    - Subprocess 3: Number of NCIs completed each year, costs, and IWC departments where conducted
- Customer and expectations:
    - IWC management (internal customers): Timely and complete investigative and NCI reports
    - IWC customers: Timely and complete investigative and NCI reports as applicable to external customers

Using this identification process, the IWC CSO can view a summary of not only investigative and NCI organizational security metrics and process related information, but the CSO can establish a form or format for such summaries and require its use throughout the security department.

SUMMARY

The investigations and NCI organization is a highly visible and important function within the IWC security department. Using a security metrics management program to assess effectiveness of the function is critical to a CSO. The information collected and analyzed can be used to improve the process and help mitigate risks, and thus better protect IWC assets.

CHAPTER 16

Government Security

CONTENTS

INTRODUCTION

When one thinks of government security and government contractors, the image of corporations that build weapons systems for the federal or central government come to mind. However, there are many other types of services and products that a government contracts for. There are corporations who engage in contracts with the government for intellectual products (information), such as "think tanks" and universities. There are also government contracts for office supplies, janitorial services, equipment, bolts and nuts, and everything in between. The preponderance of work and activity is unclassified, although a large part involves classified work.

Classified work is work that falls under the realm of national security regulations and access to classified information, areas, products, and/or equipment that are based on a security clearance (stringent vetting process) and a need-to-know (NTK) principle. That is, if you are needed to help support the contract, you have a NTK. NTK, coupled with a government-issued security clearance, is required by any individual before being granted access to facilities, data, information, or material classified under national security laws and regulations. If you are not involved in any aspects of the contract, then you do not have a NTK and access by you is denied.

Government contracts range widely in scope, product, and services. There are government contracts in the United States, for example, at the local, county, state, and federal level. They may be handled like any other contracts between a buyer and seller of products or services. However, at the federal level, there are also those contracts of a national security nature that are identified as "classified." That is because these contracts require access to or generation of classified material. The contracts are kept secure and access to classified material is based on two factors: (1) possession of a government security clearance and (2) having a NTK for related information as previously stated.

Security Metrics Management. DOI: http://dx.doi.org/10.1016/B978-0-12-804453-7.00016-1

These contracts may be publicly known with portions of them requiring access to classified information and material. Some are also known as "special access programs" or "sensitive compartmented information" programs because of their extremely and highly important national security nature. These require even more stringent access requirements and security controls. Others may not be known and are referred to as unacknowledged programs because this existence is unknown to all but those few granted access.

Government contracts carry with them certain specifications that may be universal or common and can be found in most contractual agreements between two corporations, with one being the buyer and the other the seller. However, there are contract specifications that are unique to government contracts. This is especially true if the contract is classified requiring protections under national security regulations.

If a corporation has a government contract, the contracts will often have specifications dealing with the corporation's responsibilities to protect the government assets being used and under the care of the corporation. In addition, if the contract is a classified government contract, the Chief Security Officer (CSO) is also responsible not only for the protection of the government assets related to the contract, but also providing a special security umbrella due to the classified nature of the contract.

In order for a corporation to do classified work with the US government, that corporation must enter into a security agreement with the government and will be bound by applicable government security regulation for protecting classified information and material. Failure to comply with security requirements could lead to contract default, termination, and disbarment from doing future classified work with the US government.

Some of the best assets protection practices that can help protect corporate assets can be found in the government contract environment. When you think of it, there is little difference between the two. Whether it is personnel, information, or some form of physical assets, they all require protection. Furthermore, they all require protection in an efficient and effective way. Even the reasons for the assets protection are similar:

- The loss of sensitive, corporate information may adversely affect a company causing it to lose a competitive edge or market share. If that loss is large enough it could do serious harm to the company.
- The loss of assets by a government could have serious and dire consequences, such as the defeat of the nation-state in war.
- Both have at stake the potential for a major loss of assets, leading to their demise or at least weakening them.

In the United States, the National Industrial Security Program (NISP) is the program that establishes security requirements for industry when engaged in classified contracts with the US government. The Defense Security Service (DSS) has security cognizance over all Department of Defense contractors. Other government agencies have security cognizance over their own security programs. (See the DSS web page at http://www.dss.gov for additional information.)

The US President has established the Office of Homeland Security to be the focal point homeland protection. This relatively new agency works with industry to accomplish its objectives. One may wonder if that really has anything to do with protecting corporate assets. Keep in mind much of the nation's infrastructure is in the hands of or under the control of corporations and industry. For example, the commercial communications infrastructure within the United States is largely controlled by communication corporations. Homeland Security is charged with ensuring the protection of the nation's infrastructure and, in the example of the communications infrastructure, must work with the owners—corporations—in order to effectively provide protection.

Homeland Security engages with corporations in all sectors where corporations own or operate some or the entire infrastructure. This engagement includes providing threat information, conducting security assessments, offering guidance, and any other protection action or measure that may be required or necessary. Furthermore, corporations work with Homeland Security in providing it products and services to support the homeland security mission. Much like the Department of Defense, Homeland Security relies on corporations to develop and produce the tools it needs to accomplish its mission.

Information and the nation's information infrastructure are also of paramount interest to the United States in terms of the necessity to protect. This is not unique to the United States. Every nation-state has a need to protect its own infrastructure and acts to do so (see http://www.nipc.gov/about/about.htm).

CSOs whose corporations are considered vital to their nation-state's information infrastructures have a vital role to play in their protection by protecting the assets of their corporation and thus also the information infrastructure of their nation-state.

GOVERNMENT SECURITY DRIVERS AND FLOWCHARTS

The security drivers for any classified government contract are the contract security specifications contained within the contractual document and address specific protection and security requirements. Fulfilling the contractual security requirements is essential in order to avoid such actions as increasing the potential for compromise of classified information, defaulting on the contract, and other serious consequences.

As the CSO of International Widget Corporation (IWC), the government security process is handled like all other security processes. It is important to graphically depict the macro process and related subprocesses. After all, any process analysis will begin here. Once the government security process flowchart is developed, the CSO can draw from other security functional flowcharts and embed them in the government security charts as applicable. Where the security functional processes are different, those flowcharts can be modified. This is a simpler and a more efficient method than developing the flowcharts from the beginning.

The flowchart can also be used as part of a brief to government security personnel and others who will be inspecting and auditing the performance of IWC regarding meeting contractual specifications.

GOVERNMENT SECURITY EXAMPLES OF METRICS

Assessing and managing performance within the government security program is just as important (if not more important) as in other security processes. Generally, companies doing classified business with the government spend more on security than purely commercial companies, as the provision of the NISP requires a higher degree of protection for national security (classified) material and information.

What measures are used and how performance is assessed are influenced by the provisions of the NISP and organizational needs as determined by the CSO.

Examples of security metrics in Appendix can also be very useful for a CSO in the effort to understand the government security profile (size and shape of the effort) and better help in the effort to evaluate efficiency and effectiveness. Of course, all the related security metrics data and their associated charts from the matrixed security functions would also be used as modified for use in analyses and briefings to include only specific contractual-related data and charts—e.g., personnel security: number of accessed IWC employees to each particular classified contract.

GOVERNMENT SECURITY CASE STUDY

As the IWC CSO, you have just been informed that the cognizant government security officer for one of IWC's classified contracts will be coming to the main IWC facility to conduct a security compliance inspection. What would you do to prepare for such a visit?

The IWC CSO immediately notified executive management and broadcasted an inspection notification to all managers and employees working on the contract. The CSO requested that they all conduct a self-inspection using one of the previously provided security compliance checklists for self-inspections. Furthermore, if they needed assistance or advice, they should contact the manager of the IWC government security organization. The CSO also had the security staff conduct a contractual compliance inspection of its own (security department) operations.

In addition, the CSO directed the manager of the government security organization to prepare a status briefing for the government security inspection team. That briefing would consist of the following:

- IWC's government security organizational chart and how that department is integrated into the IWC security department.
- The government security macro process flowcharts along with key micro process functional flowcharts.
- The specific security functions performed under the contract.
- The number of dedicated (direct-charge personnel time and materials) security staff supporting the contract.
- The number of indirect (overhead charges of personnel and materials) security staff supporting the contract—that is, the number of matrixed security staff supporting the contract.
- The amount of classified material by category possessed for that contract and their locations.
- The number of approved standalone and networked information systems used to process, store, display, and transmit classified and unclassified information in support of the contract.
- The location of offices and areas where contract unclassified and classified work was being conducted.
- The number and types of security containers and areas being used to store classified materials.

SUMMARY

Many corporations do business with governments. When doing business with the federal government, a corporation may be involved in classified work, which is work related to the national security and governed by legislation, regulation, and related directives. Classified contracts with the government include specific security requirements often more stringent than corporations doing only commercial business typically encounter. Adherence to the security requirements of classified contracts is critical, and failure to do so may lead to contract default or termination.

Security metrics management techniques can be used as a tool to assist a CSO to better understand his/her organizational support to the government customers and the government contracts, and how all those support processes flow—their effectiveness and costs—as well as provide that visibility to the government customers.

CHAPTER 17

Information Systems Security

CONTENTS

INTRODUCTION

Information Security (InfoSec) has an interesting history as a security function. It was developed and shaped within the information technology (IT) community and not within the security community. This was due to the fact that only the IT community had sufficient expertise to understand the information technologies being used, their applications, and their vulnerabilities. In the emerging years of IT, security organizations were not equipped with sufficient technical expertise to provide appropriate protections (beyond physical security), so the responsibility for most of today's InfoSec has stayed within the IT community.

This condition has created problems which still plague us today. One way of looking at this condition is similar to "the fox guarding the henhouse." Perhaps the best way to understand the current condition is to recognize that the service provider is also responsible for protection. When conflicts arise, delivering services is usually a priority over protection. Therefore, security—that is, protection of these valuable corporate assets—takes a back seat. What that has led to is a history of security being an afterthought, leading to preventable but frequently occurring security problems with information systems and their other related assets. Compromise of information, corruption of data integrity, denial of services, and theft of the hardware are just a few of the continuous security problems plaguing information systems. Nonsecurity IT professionals have a very poor track record of protecting these most valuable of corporate assets.

In today's information-knowledge-based society, information and those systems and networks that support information generation, processing, dissemination, and use are essential to the success of corporations. In the business world, effective and efficient information systems support a corporations' ability to meet global customers' needs. After all, information is power, today more than ever, and that can equate to a competitive edge in the marketplace.

Security Metrics Management. DOI: http://dx.doi.org/10.1016/B978-0-12-804453-7.00017-3

The security professional generally maintains responsibility for the physical protection of physical assets related to InfoSec (computers, servers, and all other related hardware). Moreover, personnel security falls within the domain of the security professional. Furthermore, within the government security function, the security professional almost always has responsibility for leading InfoSec, particularly as it involves classified information. Since the security professional traditionally manages the government security function within corporations, they generally assume responsibility for all classified InfoSec.

INFOSEC DRIVERS AND FLOWCHARTS

One of the primary drivers for the InfoSec statement of work (SOW), or security workload, is the number of information systems users driving that workload. As the number of users increases, the amount of associated security work also increases. The other primary workload driver of the InfoSec function is the number of information systems. In today's modern corporation, information devices include PDAs, notebooks, cell phones, and their associated functions of digital cameras, Internet access capabilities, and so forth.

Since information systems and users are significant drivers, the International Widget Corporation CISO, who is matrixed to the Chief Security Officer (CSO), concluded it was essential to understand and document the macro process for user activity. From that, key subprocesses that drive additional security work could be identified.

The Chief Information Security Officer (CISO), working with the CSO, developed additional InfoSec flowcharts for incorporation into both the CISO and CSO security metrics management programs (SMMPs). The CISO also decided to provide the CSO with a flowchart that mapped InfoSec support to goals.

INFOSEC CASE STUDY

The CISO was concerned about the increasingly high amount of overtime being charged by InfoSec staff members. The CISO was aware that "burn-out" could occur, which in turn leads to errors in judgment and even health problems for the staff. The CISO discussed the matter with the CSO and sought advice as to how to use a security metrics management process to collect data, analyze that data, and develop an argument for increasing the size of the InfoSec staff.

The CSO not only supported the CISO by providing advice but also by explaining that SOW time and overtime data should be collected, analyzed, and graphically depicted in an appropriate way in order to illustrate the increase in the size of the SOW (also referred to as workload).

SUMMARY

The InfoSec function has for decades now been one of the functional responsibilities of the IT staff. This is because long ago the security "professionals" abdicated their duties and responsibilities, albeit for what appeared at the time to be seemingly sound technical reasons. It is only now that security professionals recognize the need to bring InfoSec back under the overall security (assets protection) umbrella and are now trying to "get the genie back in the bottle" and under control.

As with all security functions and InfoSec is no exception, the uses of an SMMP can greatly assist in determining the costs, benefits, successes, and failures of the InfoSec function.

CHAPTER 18

Mergers, Acquisitions, or Divestitures Security

CONTENTS

INTRODUCTION

There are hundreds of mergers, acquisitions, or divestitures (MAD) taking place each year. All too often the chief security officer (CSO) and security are not part of this process or are brought into the process only during the latter stages of the effort, thus reducing their potential additional value and effectiveness. In order for the MAD to be as successful as it can be, it is essential that security be part of the process from the very early stages.

The CSO or the CSO's representative should serve as a team member or at least an "internal consultant" to the MAD team, providing guidance and direction for all security matters. In essence, they are the MAD team's subject matter "expert" for all issues of security and assets protection. In addition to the security consulting role, the CSO also must fulfill the role of a department or functional (e.g., the macro security function) manager responsible for directing and managing all security department related activities in support of the team and the overall effort.

The role of a security departmental or functional head is similar to the role of any other functional leader on the MAD team. Each is responsible to ensure they manage all activities supporting the MAD-related effort, for their specific discipline and department. For example, the CSO manages all relevant assets protection issues, just as the manager of finance ensures all financial matters are properly handled by his or her organization. Support from security for the MAD effort can be divided into two phases:

- Premerger support
- Postmerger support

Security Metrics Management. DOI: http://dx.doi.org/10.1016/B978-0-12-804453-7.00018-5

Furthermore, the contribution security makes to the entire effort can be further segmented into the following three categories:

1. *Protecting the effort itself*: Security measures are applied to the MAD effort to ensure that it is properly protected. That includes the implementation of measures to protect the confidentiality of the effort and its people, information, and physical assets.
2. *Providing subject matter expertise*: The security manager or professional serves as a member of the team and provides guidance, direction, and consultation on all security-related issues and concerns.
3. *Evaluating the security condition of the target company*: The security manager is charged with assessing the security condition of the target company. Here the CSO conducts a security survey of the target company's security program and overall security condition. That becomes part of the total team's assessment. Any issues identified are given consideration in terms of their effect on closing the deal.

As for the use of the security metrics management process in support of an MAD, it can be useful in all phases of the effort. From the task of integrating the two security organizations of two merging companies into one, to the divesting of a single business unit or significant corporate assets and the impact that will have on the security program, security metrics are used to better understand the effectiveness of the security process in support of MAD. Security metrics are also used to assess the impact of MAD on the security organization itself.

Which metrics are used will depend upon the situation. One can expect that any cost-related metrics along with efficiency metrics will be most useful. This is particularly true when combining two security organizations into one. In this consolidation process, there will be much pressure to reduce costs. Therefore, the CSO must look to find ways to reduce costs. Ideally, the CSO, using metrics management, will be able to analyze the security organization processes, keep the most effective, and change or improve the least effective, thereby creating a better security organization than either one of the two was prior to the merger.

MAD SECURITY DRIVERS, FLOWCHARTS, AND CHECKLISTS

What drives the security work for MAD? Obviously, it is governed by the corporation's executive management decisions to divest a portion of the corporation, merge with another, or acquire another business or corporation.

Remember in earlier chapters, flowcharts were used to depict how a process works—the individual steps within that process and sequence in which they occur; as well as the relationship between drivers and processes. Flowcharts and checklists can be useful tools for the CSO in executing performance and management of security processes supporting the MAD.

Since the MAD projects are not under the direct leadership of the CSO, the CSO will serve as a team member providing expertise as required by the team leader. Therefore, the CSO concentrates on developing processes to support the team's objective, one of which is the gathering of information. One important process the CSO must support in the premerger and acquisition phase is the gathering of competitive intelligence. To that end, a checklist can be very helpful in ensuring that all basic steps in the process are completed.

As part of this effort, the CSO should also establish metrics to track all tasks to include the expenditure of how many hours doing what tasks and their associated costs and materials. This will be useful in that such work should be part of a separate MAD budget and not part of the CSO's security department's budget. Use of the current security department budget should be avoided to ensure that subsequently the CSO does not go over budget. The CSO should determine how others assigned to the MAD team are budgeted for this project and ensure that the same methods are used. If each department is to use its current budget, usually an exception rather than the rule, then so be it, as long as security is treated the same as other departments.

Such data collection will also prove useful for any future MAD projects. These costs should also be tracked to each task's process and each task's process should be flowcharted, as is done with the primary security functions.

MAD—EXAMPLES OF METRICS

As stated earlier, security metrics management techniques in this book support many different security processes and functions. The security effort as part of the MAD integrated process team (IPT) draws upon these security processes to support and contribute to the success of an MAD. Moreover, the CSO must use supporting process metrics as a tool to help manage these processes as effectively as possible. Thus, for those affected processes, process metrics management is essential.

Since security metric management for the security processes used in support of MAD was discussed earlier, we will shift the focus of this chapter to the use of other security tools. This section will deal with the appropriate checklists that the International Widget Corporation (IWC) CSO will use in support of this project. Supporting an MAD is not an everyday occurrence.

Therefore, our intention is to focus in this chapter on what the CSO must do in support of an MAD so that nothing is omitted, overlooked, or forgotten. As part of the checklists, questions will be asked which should drive the CSO to develop measures unique to MAD and useful in assessing security performance.

CHECKLISTS

Premerger and acquisition checklist: During the pre-MAD phase, the high-level objectives for the security manager are to protect the effort and to participate in the due diligence. During this phase, the CSO has many distinct tasks to perform that contribute to the overall success of the effort. Checklists below will be useful to security managers and MAD team leaders as they develop their plans.

Protect the effort: Develop an Operations Security plan that includes guidance and requirements for the following areas:

- Maintain confidentiality of the effort's existence and activities
  - Nondisclosure agreements to be executed by all participants
  - No public release without team leader approval
  - Identification of designated project areas.
- Confidentiality of the team's work: Brief team members on:
  - Need for strict compliance with need-to-know
  - Requirements for protecting information
  - General procedures, requirements, and other obligations.
- Protection of information
  - Define information sensitivity—what must be protected and why.
  - Identify Information Administrator—control and tracking of project documents.
  - Create an information control plan addressing the following areas:
    - Handling of information (paper and electronic)
    - Creation of documents (paper and electronic)
    - Distribution of documents (paper and electronic)
    - Electronic transmission of information and documents
    - Reproduction of information and documents (paper and electronic)
    - Destruction of documents (paper and electronic).
- Protection of all MAD team members
  - Conduct general threat and vulnerability assessment
  - Identify all team executives and members
  - Prepare for international travel
  - Obtain country risk assessments
  - Provide team members with travel safety briefings
  - Develop and brief the cover story for all team members to use.

Due diligence: During the due diligence effort, the security manager has two primary goals:

- The first is continuing to protect the effort, particularly team members as they move from location to location.
- The second is to conduct the security portion of the assessment on the target company.

  The results of the due diligence assessment will be used by the chief executive officer (CEO) and/or board of directors to assist in making decisions relative to proceeding or not proceeding with the merger or acquisition.
- Competitive intelligence assessments: Security must be a member of the competitive intelligence team providing information and investigative support for the following assessments:
  - Market sector
  - Target companies
  - Competition intentions.
- Background investigations are to be conducted as needed on the following:
  - Target company executives and board members
  - Target company affiliates and associates
  - Target company key supplier executives.
- Evaluation of the target company's security profile
  - Assess current security condition
  - Collection of data for organizational analysis
  - Identify potential operational synergies
  - Identify any major vulnerability
  - Identify any deal-breaking issues.

Postmerger and acquisition checklist: The major activity of the post-MAD period is combining organizations. The newly acquired company or business unit must be integrated into the acquiring company. Each functional leader must participate. The CSO is responsible for integrating the security department from the acquired business unit into the acquiring company. In the process, there are tasks to accomplish and issues to address:

- Develop plan to combine security organizations
  - Leadership: A single leader needs to be recognized and tasked.
  - Create a single security department.
  - Establish common vision, operating philosophy, goals, and objectives.
  - Analyze and define new security statement of work.
- Security operations
  - Implement a common operation—common and consistent application of security services and requirements is essential.

  - Process—analyze all processes—adopt the best, improve the others.
  - Operational synergies—identify areas where savings can be achieved.
  - Best practices—identify practices from both companies and adopt the best, and discard or improve those that are inefficient.
  - Policies and procedures—review all policies and procedures. Ensure they are revised to properly reflect changes made during integration and take advantage of the best both companies have to offer.
- Budget analysis
  - Staffing requirements—aligned with the security statement of work?
  - Reorganization—to ensure security professionals are best matched with security tasks.
  - Cost control—identify ways to improve the efficiency of department processes and procedures.
- Company-wide security issues
  - Identity issues—badges for all new employees.
  - Vetting of new employees—for target company employees never before subjected to a preemployment or suitability-for-hire background check—consider doing.
  - Work Place Violence prevention—ensure a prevention program is in place, staffed, and all intervention team members trained.
  - Customer notifications—as needed.

Divestiture checklist: When a business unit is divested, the CSO has two primary actions to take in support of the activity. One is a protective effort while the other is a management action in separating out those security systems and employees that go with the divested unit. The protective effort could include providing support to the divested business unit should the company decide that is in its best interest.

- Separation of the divested business unit
  - Identification of employees badges—collect company badges from departing employees. Work with acquiring company to establish new identity process.
  - Access controls—ensure departing employees have access only to that information and those areas for which they have a need-to-know.
  - Cost containment—ensure separation occurs as swiftly as possible. Delays translate to added costs and expenditures.
  - Ensure only those assets defined as part of the deal go with the divested company. Anything not part of the deal, stays.
- Security Infrastructure
  - Systems—separate the departing business unit from the security systems being used.
  - Alarm systems—remove from divested buildings and facilities.
  - Access control systems—reconfigure systems to accommodate operational changes.

  - Security information—remove divested employees from the security information systems database. Transfer data to acquiring company (if appropriate).
- Employees—be prepared for employee issues in the following areas:
  - Unauthorized removal of assets and information
  - Potential for workplace violence
  - Seeking general security guidance about rapidly changing events
  - Maintaining regular communication with employees advising them of changes, security concerns, and events.

Lessons learned checklist: Many companies choose to shape their strategic direction through the use of MAD. The likelihood of MAD occurring more than once in any company is increasing. Therefore, it is a prudent action to learn from one transaction in order to better prepare for the next. Below are areas to consider when reviewing lessons learned:

- Use of Competitive Intelligence
  - Did the competitive intelligence produced add value?
  - Did it increase team knowledge of the target company?
  - Did it increase team knowledge of market?
  - Did the intelligence contribute to the team's success?
- Was the effort properly protected?
  - Were there any security failures?
  - Was the effort exposed to anyone without a need-to-know?
  - Were any team members exposed, embarrassed, or harmed?
- Assessment of target
  - Was the target company properly assessed?
  - After the deal was closed, was anything significant about the acquired company learned that could/should have been learned earlier?
  - Did the assessment lead to a realistic valuation?
  - What problems developed?
- Integration success
  - Were the goals and objectives of the team, the CEO, and the board of directors met?
  - Did the team perform as expected?
  - Were costs objectives met or exceeded?
  - Did any issues of culture develop—i.e., culture clashes?

MAD CASE STUDY

The MAD team has never been involved in such a project before and the team leader has asked the CSO to explain in business terms how the competitive intelligence effort should be handled.

The IWC CSO, an experienced CSO, has always worked to cultivate sources of information that can assist the CSO in protecting the corporate assets. These same sources will prove useful in assisting the CSO support the MAD project team's competitive intelligence collection efforts.

It was decided by the MAD team that one of the primary functions of the CSO as part of the MAD project team would be to lead an effort to gather competitive intelligence on the corporation that has been targeted for acquisition. The CSO developed a simple chart in order to explain to the MAD team the basic philosophy that the CSO uses for such projects.

This is an important function and often accomplished by nonsecurity personnel. However, often the CSO is in the best position to provide this information or at least be part of the project team members responsible for gathering the information. Some corporations have a competitive intelligence organization that specializes in gathering information about its competitors. If so, the CSO should already have opened communication channels with this group and, as part of an MAD, assist them.

SUMMARY

MAD are strategic business tools used more frequently in this expanding global marketplace, to enhance the competitive position of the corporation. The CSO must play an important part as an integral member of the MAD team. As such, the CSO has certain responsibilities in the pre-MAD and post-MAD phases. The use of a security metrics management program can greatly assist the CSO in successfully accomplishing the tasks assigned by the MAD team.

CHAPTER 19

Outsourcing

CONTENTS

INTRODUCTION

Outsourcing, by which we mean "contracting for outside services that are a necessary part of doing business but are not core functions or core competencies," is always an option when a Chief Security Officer (CSO) is looking for ways to save money for the corporation and cut costs for the CSO's security department.

There are many advantages and disadvantages to outsourcing a security function. As a CSO, you should look at all security functions as possibilities for outsourcing, as long as they are not considered "core competencies" by the corporation—and usually they are not.

When considering outsourcing, the CSO should first look at transactional processes (generally the same steps occur within each process transaction) such as preemployment background investigations and other security processes where there are many capable providers of cost-effective services (such as security guard services). More complex processes such as information systems security are less desirable targets for outsourcing. Outsourcing processes that require a high degree of skill and training means the company no longer has that intellectual capability or asset in-house. This can be a precarious position as there is little to hold or bind the skilled (outsourced) provider to your corporation other than money.

As a CSO, you more than likely know other corporate CSOs who have been involved in outsourcing some of their security functions. If so, they should be contacted to ascertain what lessons they learned during the process and how the process currently works—are they saving the money they thought they would save and still getting satisfactory service?

It is interesting to note that the estimated savings expected from outsourcing does not always occur. One reason for this is optimistic savings projections

Security Metrics Management. DOI: http://dx.doi.org/10.1016/B978-0-12-804453-7.00019-7

for outsourcing. Perhaps the cost–benefit analysis was flawed or over-optimistically skewed. Having a security metrics management program in place, especially a mature program with several years of tracking costs, will help reduce the likelihood of a flawed outsourcing cost–benefit analysis.

As a CSO considering outsourcing a security function, you should seek answers to the following basic outsourcing questions from other CSOs who have outsourced a security function that you are now considering for outsourcing:

- What processes did they outsource?
- What outsource provider did they use?
- What others were considered or not considered, and why?
- Were the expected cost savings achieved?
- If not, did they achieve any cost savings at all?
- Was there degradation in service?
- What other benefits did they derive from outsourcing?
- What pitfalls were encountered?
- Are they a satisfied customer?

The information the CSO obtains from this process will be useful in setting the CSO on a successful path. Learning from the experiences, successes, and failures of others can save considerable time and money. Once the CSO has determined candidate security functions for outsourcing, the CSO should also consider the following:

- The processes considered for outsourcing are not core competencies.
- There are no proprietary issues to be concerned with.
- Company culture or union issues do not exist (or have been reasonably addressed and they will not prevent the CSO from proceeding).
- The CSO has learned from the experiences of other CSOs.

OUTSOURCING DRIVERS AND FLOWCHARTS

The primary driver for outsourcing is of course cost savings.

Once a contract has been awarded to a service provider to perform a security function or deliver a security service, the CSO then transitions into an oversight role to ensure the provider does perform to stated requirements, providing the necessary assets protection service or process at the established cost. The data collection process used while the security function was performed in-house can be used to collect data from the supplier. Thus begins the security metrics management process (costs to be collected and delivered). However, in order to do so, the data to be delivered by the supplier must be a specification of the contract. If not, then you are asking for work to be accomplished that is not part of the contractual agreement. That may add to the cost of the effort. The supplier might so state or possibly just send an additional bill for "services rendered."

As a CSO, you may have obligated the corporation and the additional costs must be paid. So, as the CSO, you should be very clear as to what is in the contract and if you are not clear about it, ask your contracting officer (or procurement official) who is overseeing the contract for you and the corporation. If the data you requested is important and you want it, then a change in contract is the proper way to handle that request. However, be cautious here that you really need the data. You may end up paying more for the outsourced security service than if it were done in-house, at least more than originally expected, if the additional work is not well defined and is excessive.

OUTSOURCING EXAMPLES OF METRICS

After appropriate research and data collection has been completed, the CSO can then determine if it is time to proceed further with the analysis for finding a capable provider of services. The International Widget Corporation (IWC) CSO determined that the IWC procurement office may also be of assistance in identifying potential suppliers. It is necessary to identify a sufficient number of suppliers to hold a true competitive bid process. Usually that means including between three and five (or more but should not be less than three) potential providers. The more potential suppliers included, the better the competition.

The competitive bid process is the best way to ensure the CSO and IWC get the best results. The competitive bid process should provide for a fair assessment of each supplier's capabilities and ability to perform. The process should assess the following:

- Potential suppliers' experience
- Contact references provided
- Response to the security statement of work (SOW)
- Cost and affordability
- Availability for service (time and location limitations if any)
- Responsiveness and flexibility
- Available resources to perform
- Quality and metrics program

The competitive bid process should also allow each potential outsource provider an opportunity to learn more about IWC's security SOW. This can be accomplished through a conference for all bidders and/or a question-and-answer process. The bid process should also include developing a transition plan—for example, one that addresses how the outsource provider will transition from the current proprietary guard force to the contract force. The transition plan can be planned as a major milestone in the outsourcing project plan. As such, it is also to be used as a form of security metrics management.

As part of that outsourcing process, the CSO should use the security metrics management data that has hopefully been collected over the years to help not only in developing a specific SOW but also in knowing the average costs for completing that SOW, which can be used for comparison with the bids received. After all, the bids should always come in lower than doing the work in-house, unless there are some other reasons to outsource the work other than cost savings.

As the CSO proceeds with the cost analysis, the following data (e.g., security guards) should be collected from the proprietary operation for comparison with data provided by potential outsource providers, as applicable and based on the security function outsourced:

1. Budget Category
   - Total department salary by employee
   - Total bonus costs
   - Calculate cost of benefit package for all employees (current and future)
   - Calculate the cost per employee (current and future)
   - Calculate costs by security guard posts to be filled
2. Facilities Category
   - Calculate office space (sq. ft.) occupied by department
   - Calculate cost of that space for the department
   - Cost of department-owned equipment
   - Cost of common use equipment
3. Travel Category
   - Cost of all business travel and associated expenses
4. Training/Development
   - Cost for seminars/courses
   - Cost for internal meetings
   - Cost for other activities
5. Oversight Obligations
   - Cost of time demands placed on line units
   - Development and maintenance of performance measures
   - Cost of reports
6. Other
   - Legal (factor in potential claims of various types)
   - Administrative support
7. Subcontractor Category (if applicable)
   - Total cost of subcontractors to include labor, materials, etc.

A comparison of the data collected from the proprietary operation with data provided by a potential outsource provider should offer an immediate indication as to any cost differences. Placing the data into a spreadsheet listing tasks and costs in time and materials can easily show the variances by line item and in total.

POST-CONTRACT AWARD

Once a contract is outsourced, a security survey/questionnaire may be considered to determine if the IWC internal customers are satisfied with the supplier's services. This is a process for measuring customer satisfaction. Customer satisfaction is an important metric. It does not indicate efficiency, but it certainly does let the CSO know how security's customers feel about the service or products they receive. After all, security is a service and support department and as such must satisfy its customers; therefore, service and support quality is of utmost importance.

The following is a sample survey questionnaire used for outsourcing the IWC access badge process (it can also be used for in-house evaluation).

A. How satisfied are you that your employee badge was issued to you promptly?
1) Very Satisfied 2) Satisfied 3) Unsatisfied 4) Very Unsatisfied

B. The security clerk was able to answer all of your badging related questions.
1) Very Satisfied 2) Satisfied 3) Unsatisfied 4) Very Unsatisfied

In a full survey, all responses can be scored, averaged, and a baseline established. This baseline should be identified as the level of service provided going into an outsource condition. If the service was poor, the CSO's expectation may be for the outsource provider to deliver better service. If the baseline is high (very satisfied), for the sake of keeping costs low, the CSO may be satisfied with a lower level of service (satisfied), but maybe the customers would not be. Then what would you do?

It is important that services and support satisfaction measurements be identified as a specification of any outsourced contracts.

Once the CSO has made the decision to outsource and has awarded a contract to a selected provider, there are still a few security metrics related factors that should be mutually agreed to, thus enabling complete success. Doing this will ensure both parties—the customer and the provider—benefit from the newly formed relationship. The areas should be defined, agreed upon, and made part of the contract or process:

- *Identify critical success factors*—Those factors that must be fully achieved, for the effort to be considered a success. For example, if the reason for outsourcing is to reduce cost by 10% or more, then achieving that 10% cost reduction is a critical success factor.
- *Establish goals*—Clearly defined goals are a means of defining expectations.
- *Measure performance to goals*—How well the provider performs to established goals is a measurement of success. Moreover, it is a futile effort if goals are established but performance to them is not measured.
- *Improve*—Performance measurement is a management tool. It establishes where IWC is, as compared to where it expects to be. If performance is below expectations, improvements must be made.
- *Periodic reevaluations*—All relationships require periodic reevaluations. Neither customer nor supplier should assume all is well unless that is validated through some established means. For example, after doing business with a single supplier for several years, the CSO decides to recompete a contract let to that supplier years earlier. Through a competitive bid process, a CSO can validate that the current supplier offers the best deal or select another supplier who does.

OUTSOURCING CASE STUDY

An experienced CSO decided that it might be best for the organization to outsource the security education and awareness program (SEAP). Several months after outsourcing the SEAP, the CSO was asked to share her experiences on this outsourcing effort. The following is a summary of her experience:

- What made you decide to outsource? *I began to realize that the cost of providing a company-wide SEATP with internal personnel was high and we were not getting all that was needed. I needed a lower-cost program that was cost-effective and still remained interesting and fun for all employees.*
- Did you do a cost–benefit analysis as part of that determination? *I did do a cost–benefit analysis that was sold to the executive staff. Fortunately, as part of my metric management program I had cost trend data for the past 5 years. I was able to use that data and compare it with data provided by experienced service providers. I concluded I could get a better product/service at a lower cost by going outside the company and using an experienced provider. Moreover, with an in-house program, I did not have the flexibility of having the many different skills available for use that a company with a core-competency in*

SEATP has. I had a security specialist and myself. The service provider I selected has a cadre of experts to draw from.

- Has it worked out as expected? *Better than expected. We have a more flexible and efficient program provided at lower costs.*
- What specific services are provided and how are they provided, such as lectures, pamphlets, etc.? *I receive monthly newsletters, weekly news wire clips to send out to the entire organization, a dedicated artist for postings, calendar art as well as newsletter art. I have bi-monthly training, with outside speakers, as well as an entire "Security 101" training day.*
- How did you go about looking for a supplier? *I wanted someone a little "out there" and Winn [Schwartau] was mentioned as a good candidate, along with three other providers. We went through a competitive bidding process, working with the company procurement officer, and we selected Winn. We've been working together ever since.*

SUMMARY

In today's business environment there are many providers of a broad range of security services. These providers range from local companies providing few specialized services to international corporations offering security services from the simple to the very complex. A skilled CSO will recognize this and use outsourcing as a tool to help effectively and efficiently manage the security functions.

Never forget that corporate security is part of the business enterprise. Corporate security adds value to a corporation in many ways, but does not usually generate revenue. Therefore, corporate security is a cost center. Keeping costs under control and as low as possible is essential. Outsourcing provides an effective avenue for the CSO to reduce costs or keep them under control.

The security metrics management tool used for data collection, analyses, and charting provides the necessary baseline to support a project looking at outsourcing-specific security functions.

CHAPTER 20

A Look into the Future

CONTENTS

INTRODUCTION

In this final chapter, we look at possibilities of the future of the security profession and using security met7rics to support maximizing efficiencies and effectiveness of the assets protection program.

CHIEF SECURITY OFFICER PROFESSION

Looking back over the decades, this profession has been gradually changing from one where the Chief Security Officer (CSO) was a person who retired from law enforcement (local, state, or federal) to one who has been in corporate security as a professional after receiving a degree in assets protection.

In the past, the management of corporations thought hiring some FBI agent or police officer was the best way to protect assets; thus, showing their ignorance of what was required. Someone who spent their first 20 or so years conducting investigations has little in common for managing an assets protection program. Luckily for corporations, although that hiring mentality still continues albeit less than in the past, those days are numbered. Gone are the days where the CSO's responsibilities were that of managing physical security consisting of a guard force, fences, locks, and alarms. Yes, that is still required; however, it is no longer the end all, be all of assets protection.

For decades, the American Society for Industrial Security has led the charge to establish a baseline for assets protection professionals and provide a certification (Certified Protection Professional) to provide a pool of professionals for corporate management to choose from.

Security Metrics Management. DOI: http://dx.doi.org/10.1016/B978-0-12-804453-7.00020-3

CSOs are now faced with more and more sophisticated threats to assets they are to protect. Furthermore, these threats are more and more becoming international. At the same time, the CSO must also be able to safeguard the assets from the simple internal and non-sophisticated thefts and frauds.

One just has to look at the chapter titles presented in this book and also our upcoming second edition corporate security book (to be published in May 2017) to see the expanding responsibilities of a CSO. These responsibilities and processes more and more are becoming technology based.

NEW TECHNOLOGY

One does not have to look too far in the past to see the rapid changes in technology from vacuum tubes, transistors, microchips to today's small wireless devices. These mobile devices have offered new challenges to the CSO from theft of these devices with terabytes of sensitive corporate information to the hacking and theft of the information contained in these devices.

The assets—powerful, mobile devices—enable corporations to operate and compete globally, more effectively, and efficiently. They will not be going away but increase in numbers, power, and decrease in size.

In the future, the use of artificial intelligence, robotics, and miniaturization of devices will offer new challenges to the CSO in the quest to provide effective and efficient assets protection. How will the CSO deal with nanodevices, for example?

APPLYING HIGH TECHNOLOGY TO THE SECURITY METRICS MANAGEMENT PROGRAM

When the CSO looks at the future technology and its application, threats, vulnerabilities, and risks to assets protection, it should also be looked at as a tool to providing more effective and efficient ways of collecting metrics data.

The integration of the metrics processes into the automated systems used by the corporation will allow the collection of information in an automated way. Thus providing not only a nonmanual and therefore a more accurate way of collecting information but also eliminate the need for staff's time to be spent on non-assets protection functions.

EVALUATING CURRENT AND FUTURE DATA COLLECTION NEEDS

In the meantime, the CSO should continue to review, analyze, and improve on the security metrics management program (SMMP) and evaluate the SMMP periodically for relevance and utility of each measure:

- Are they meaningful?
- Do they provide an understanding or insight into the security process or function?
- Do they contribute to the understanding of the effectiveness, efficiency, success, or failure of security processes and the security program?
- And, perhaps most importantly, do they help the CSO make needed changes and decisions?

These are the questions each CSO and security professional should ask and answer to ensure that the SMMP is successfully supporting the CSO as it was intended to do.

SUMMARY

The security profession is evolving from the days when retired law enforcement officers became CSO with only responsibility for guards and physical security to one requiring a totally new profession.

As technology evolves, the CSO should not only look to its threats, vulnerabilities, and risks to corporate assets but also as tool to provide more effective and efficient assets portion program. In addition, technology should be evaluated for modernizing and integrating into an SMMP.

Appendix: Figures, Charts, Graphs, and Surveys

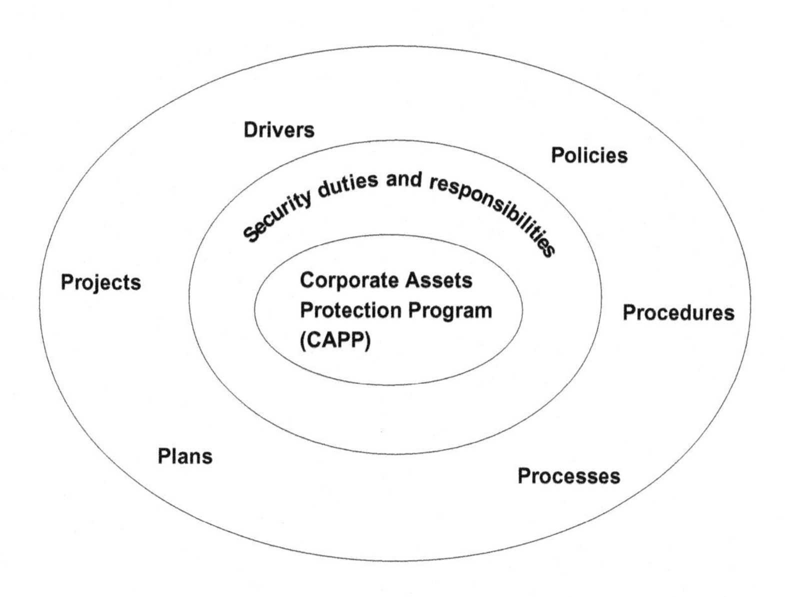

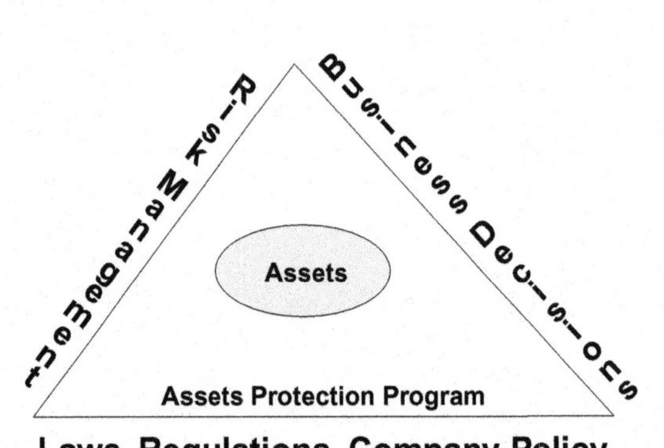
Risk Management
Business Decisions
Assets
Assets Protection Program
Laws, Regulations, Company Policy and Procedures, Ethics, Privacy, and Good Business Practices

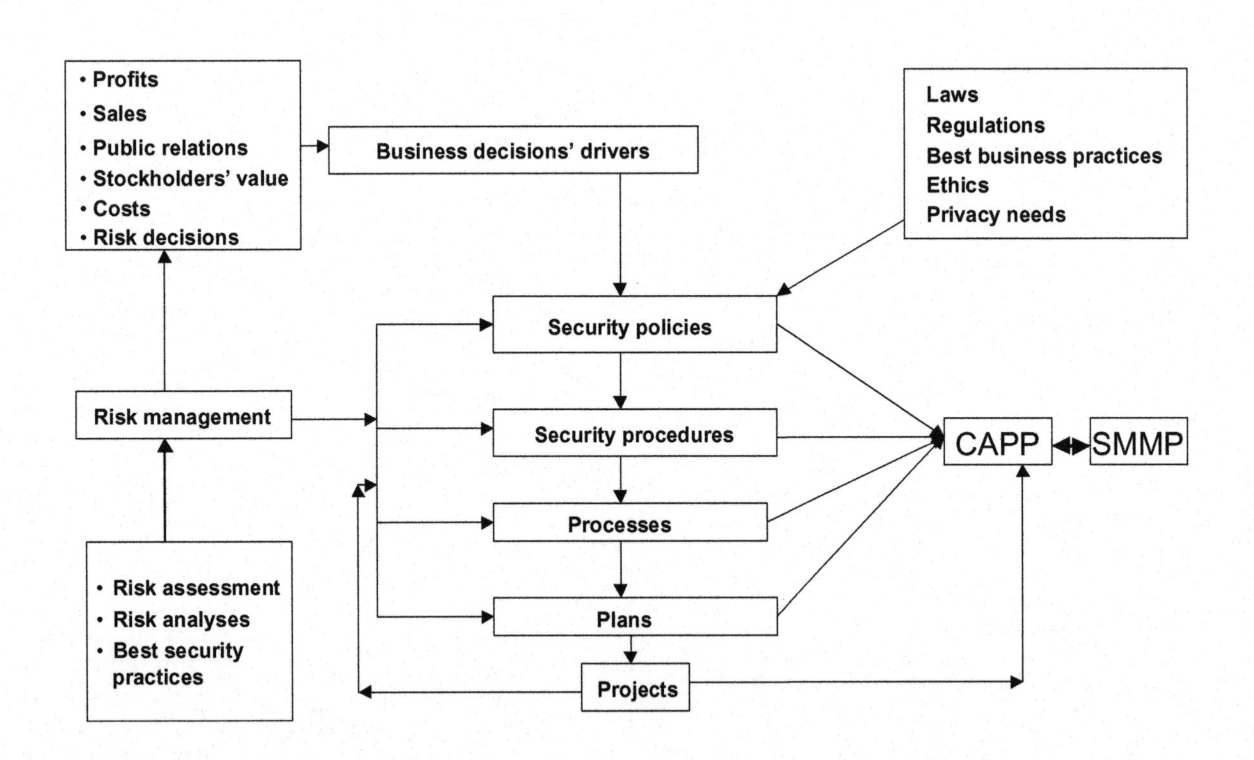
• Profits
• Sales
• Public relations
• Stockholders' value
• Costs
• Risk decisions
Business decisions' drivers
Laws
Regulations
Best business practices
Ethics
Privacy needs
Security policies
Risk management
Security procedures
CAPP
SMMP
Processes
Plans
Projects
• Risk assessment
• Risk analyses
• Best security practices

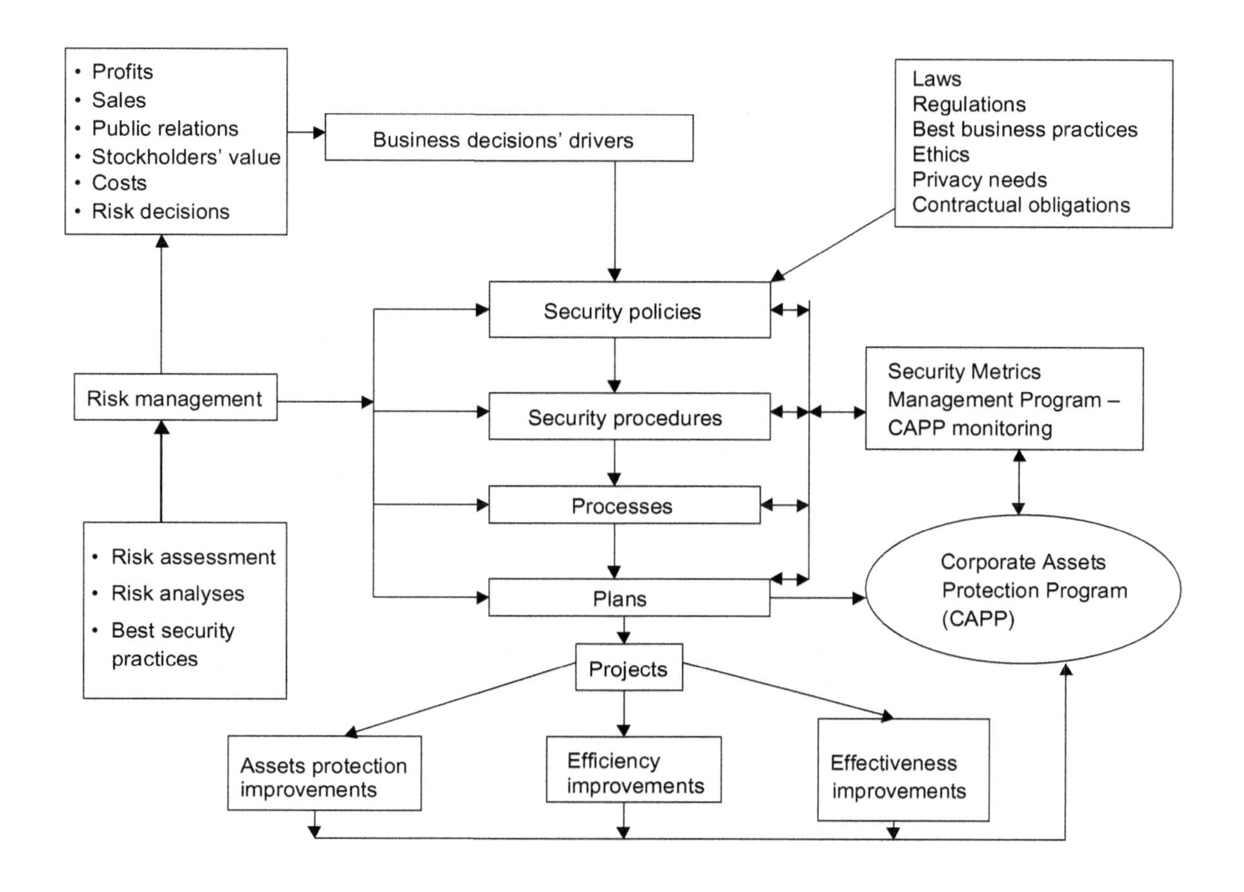

What is a process?

- A series of steps or actions that produce something

- A process statement should describe an action, not be a job, task or program/process title or description (e.g., "produce tax report," not "tax reports")

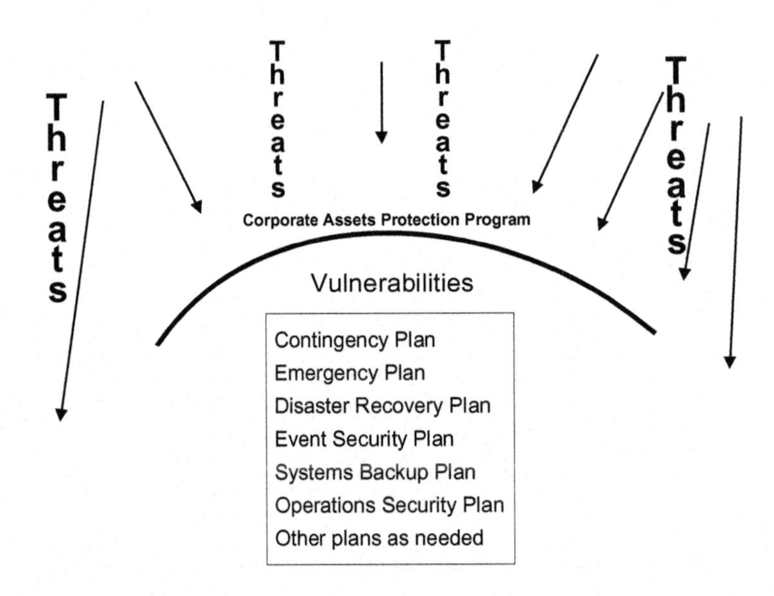

SUBJECT: ______________________
RESPONSIBILITY: ______________________

ACTION ITEM: ______________________
REFERENCES : ______________________

- **OBJECTIVE(S)**
- **RISK/STATUS:**

	ACTIVITY/EVENT	RESPONS-IBILITY	2002 Feb	Mar	Apr	May	Jun	Jul	RISK LVL	DESCRIPTION
1										
2										
3										
4										
5										
6										
7										
8										
9										
10										

Legend	Scheduled	Complete	Risk Level	
Milestone ○ =Major			HIGH	H
Activity			MEDIUM	M
Deviation Span			LOW	L

PAGE 1 of 1

ISSUE DATE XXXXXXXX

STATUS DATE ______________________

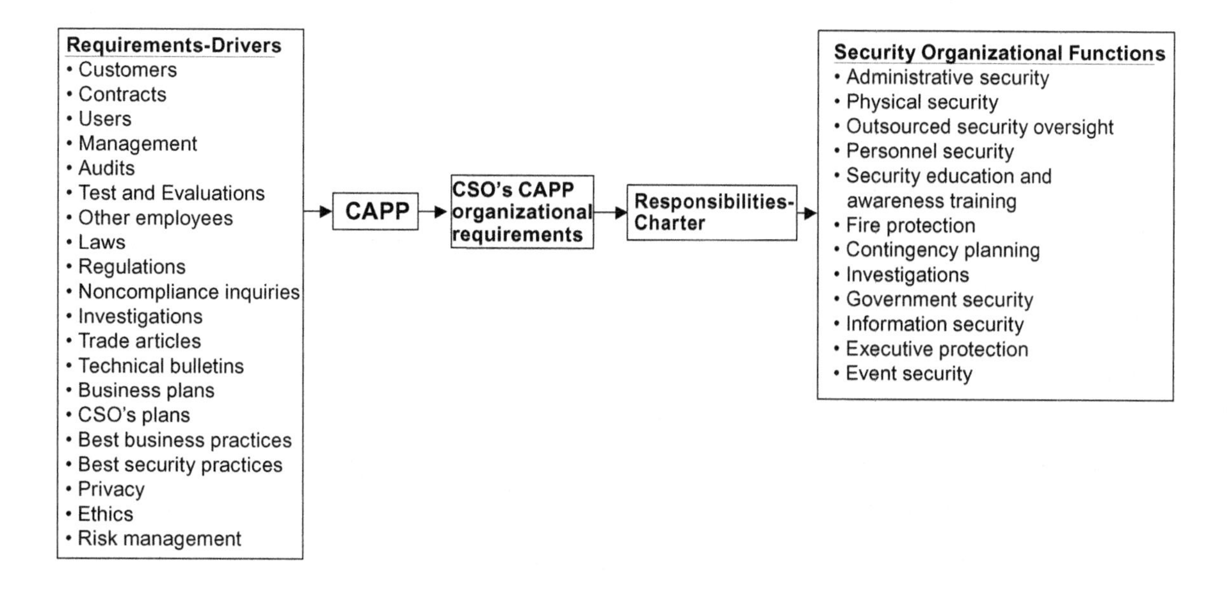

Drivers → Corporate Assets Protection Program → Security Department → Security Functions and Processes

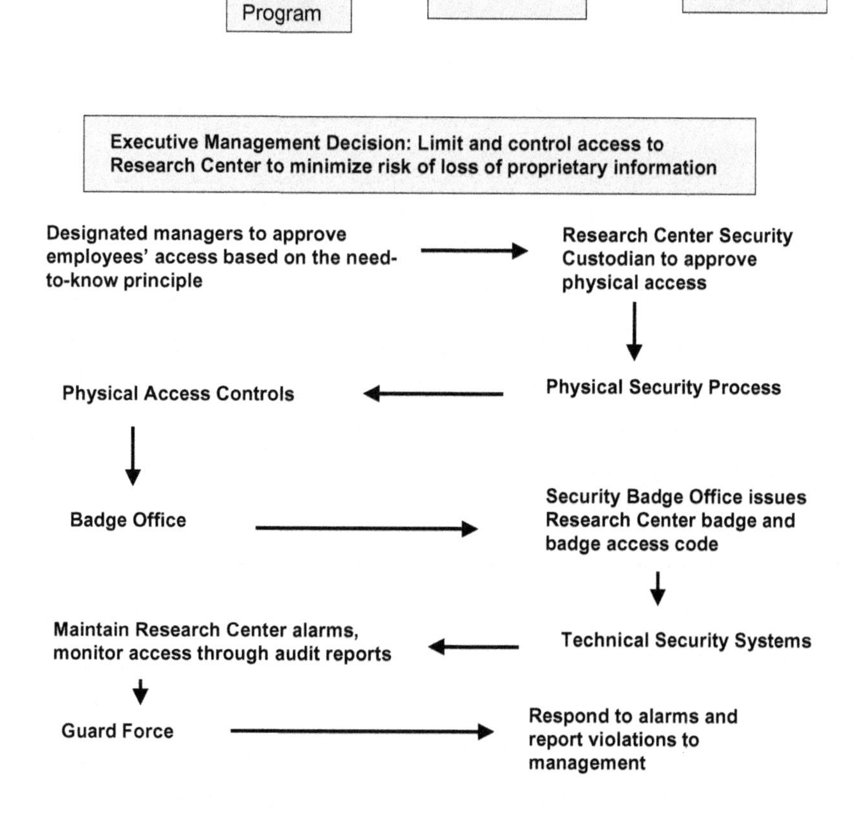

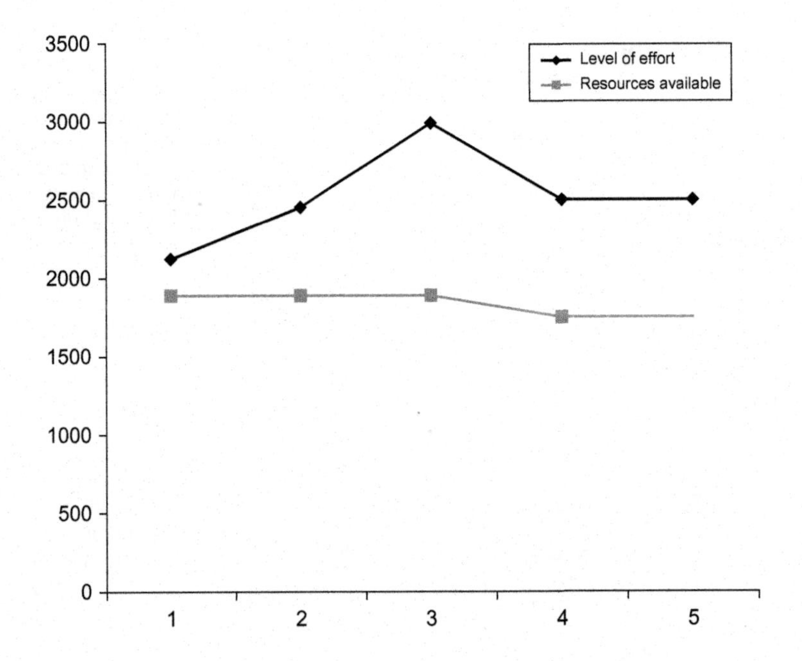
3500
3000
2500
2000
1500
1000
500
0
1
2
3
4
5
Level of effort
Resources available

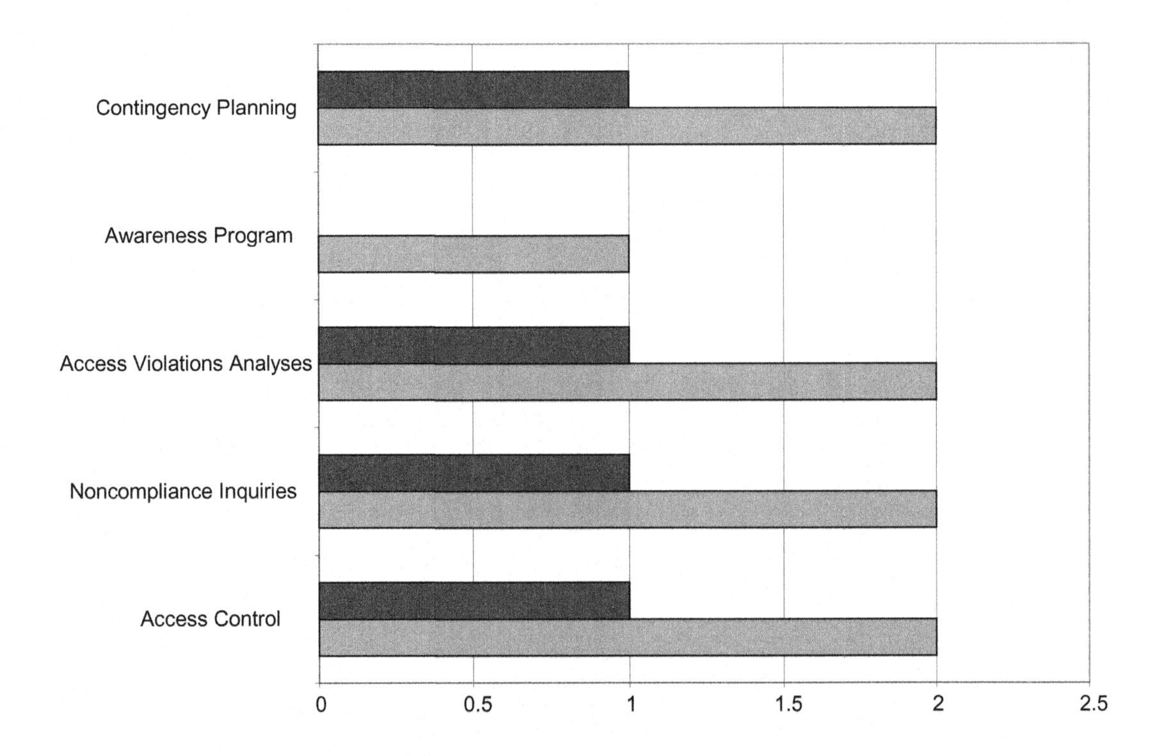

Security functional process summary

1. *Process name*: investigations and NCIs
2. *Process description*: provide professional investigative and NCI support to all the corporate sites
3. *Supplier*: customers and employees
4. *Input*: complaints, allegations, requests for assistance, asset protection requirements
5. *Subprocesses*: investigations, NCIs, crime prevention surveys
6. *Customers*: management
7. *Output*: investigative reports, security assessment reports, briefings, survey reports, testimony
8. *Requirements and directives that govern the process (drivers)*: CAPP, policies, laws, regulations, management decisions

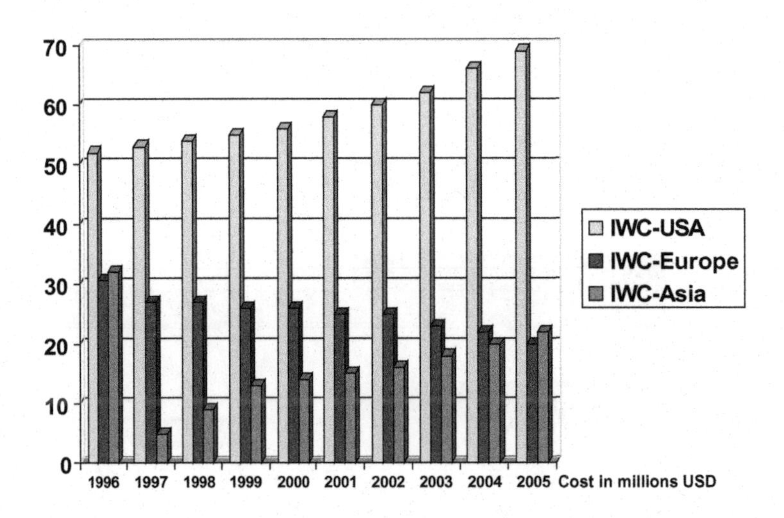
IWC Security Costs—All Locations 1996–2005
70
60
50
40
30
20
10
0
1996 1997 1998 1999 2000 2001 2002 2003 2004 2005 Cost in millions USD
IWC-USA
IWC-Europe
IWC-Asia

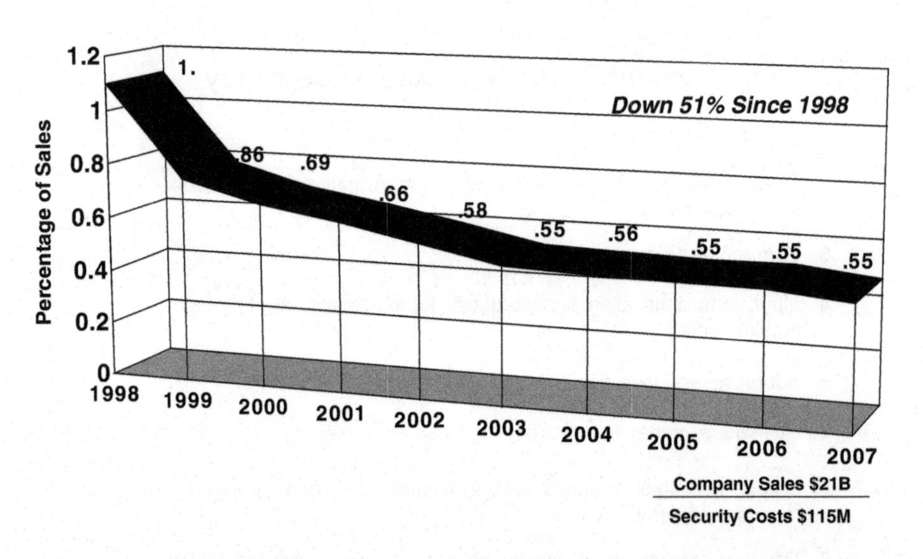
10 Year Trend
Security Costs as a Percentage Of Sales
Percentage of Sales
1.2
1
0.8
0.6
0.4
0.2
0
1.
.86
.69
.66
.58
.55
.56
.55
.55
.55
Down 51% Since 1998
1998 1999 2000 2001 2002 2003 2004 2005 2006 2007
Company Sales $21B
Security Costs $115M

IWC-Wide Assets Protection Costs—Per Employee 1996–2005

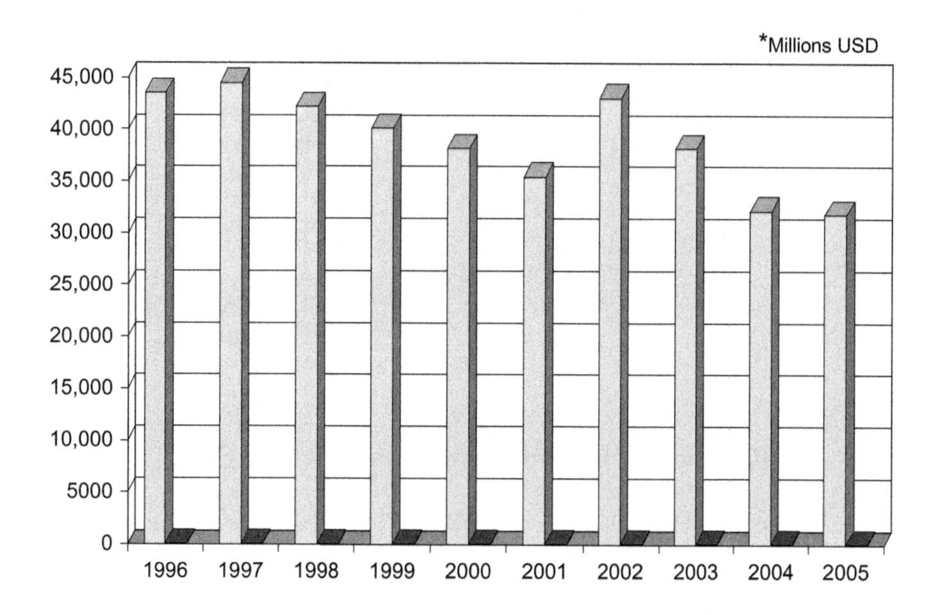

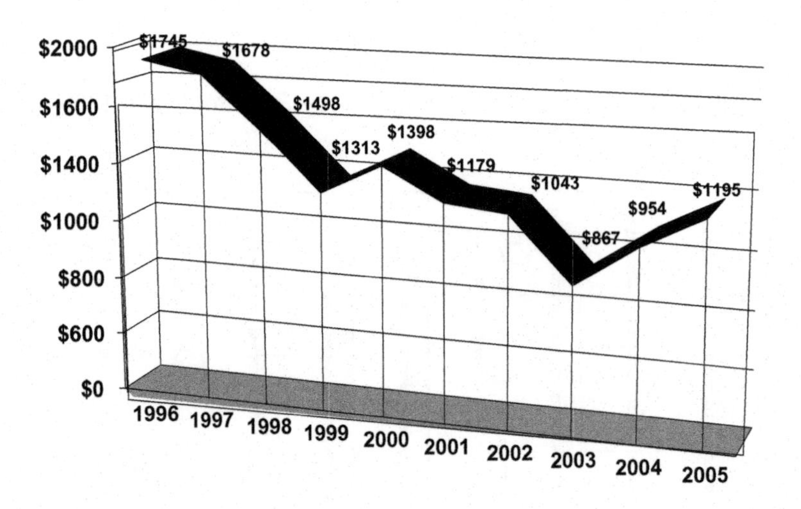
IWC-Wide Security Cost Per Company Employee
$2000
$1600
$1400
$1000
$800
$600
$0
$1745
$1678
$1498
$1313
$1398
$1179
$1043
$867
$954
$1195
1996
1997
1998
1999
2000
2001
2002
2003
2004
2005

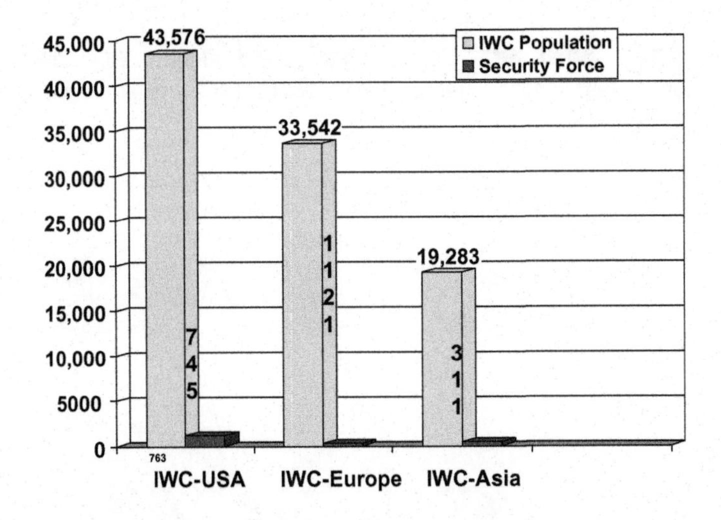
IWC Ratios by Region – Employees/Security Staff
45,000
40,000
35,000
30,000
25,000
20,000
15,000
10,000
5000
0
43,576
33,542
19,283
745
1121
311
763
IWC Population
Security Force
IWC-USA
IWC-Europe
IWC-Asia

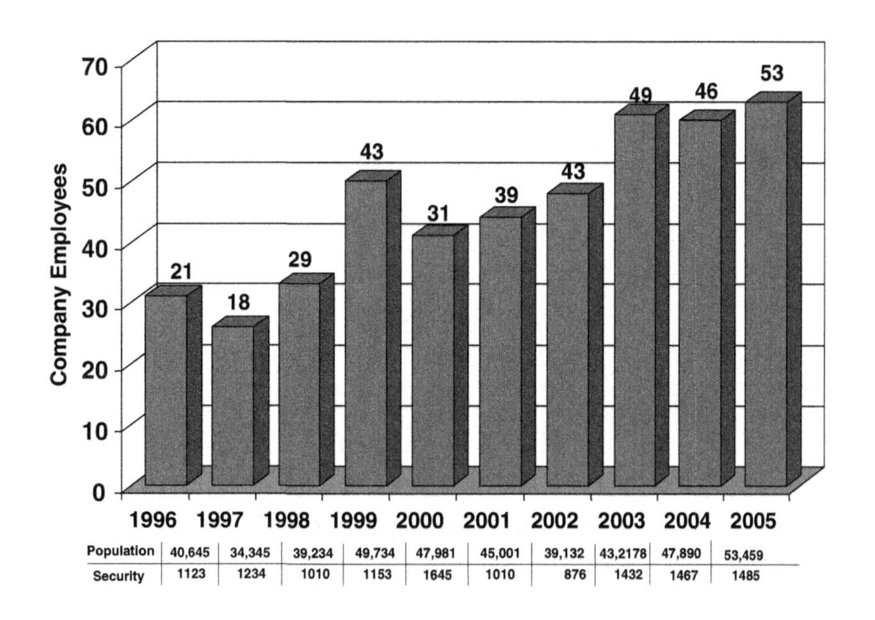

	1996	1997	1998	1999	2000	2001	2002	2003	2004	2005
Population	40,645	34,345	39,234	49,734	47,981	45,001	39,132	43,2178	47,890	53,459
Security	1123	1234	1010	1153	1645	1010	876	1432	1467	1485

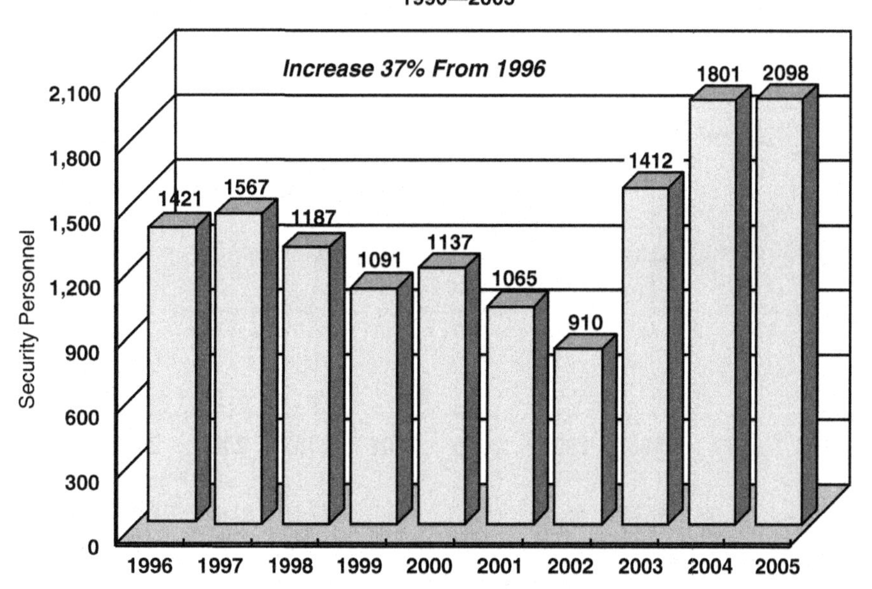

Composition of the IWC Security Workforce

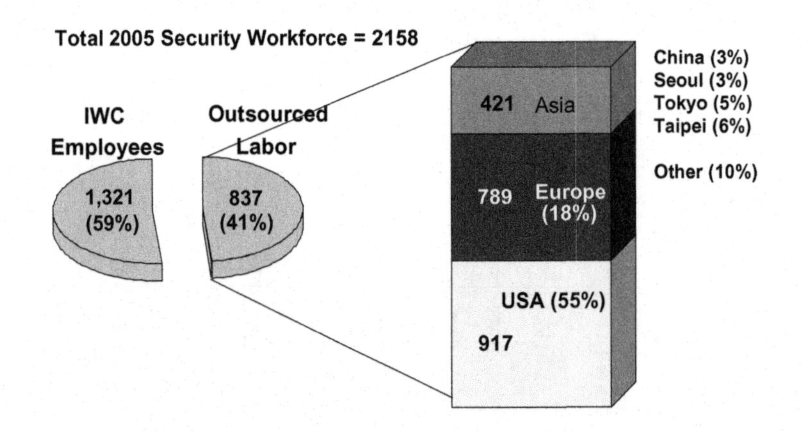

10 Year Trend
IWC Population / Security Workforce

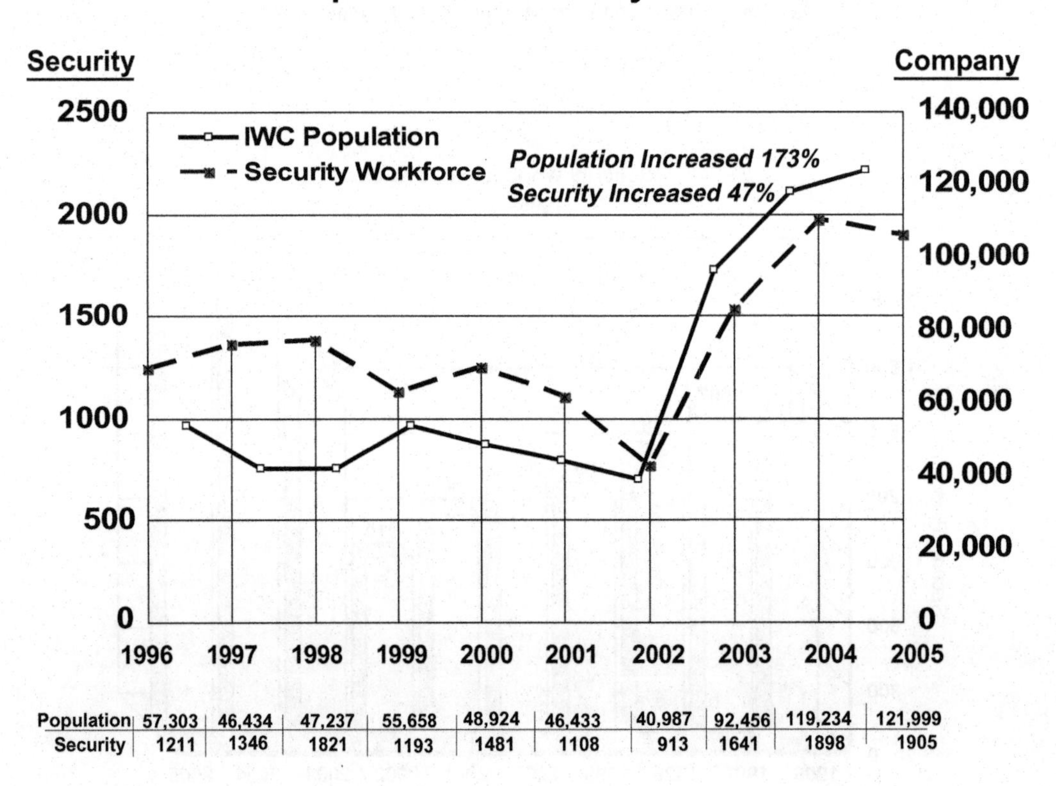

	1996	1997	1998	1999	2000	2001	2002	2003	2004	2005
Population	57,303	46,434	47,237	55,658	48,924	46,433	40,987	92,456	119,234	121,999
Security	1211	1346	1821	1193	1481	1108	913	1641	1898	1905

IWC Facility Utilization Profile in Terms of Proprietary and Sensitive Research Activity

2005 IWC Sq Ft
567,437

Sensitive
65,546
(3%)

Proprietary
43,456
(3.5%)

Employee Population & Trade Secret Clearance-Access 1996–2003

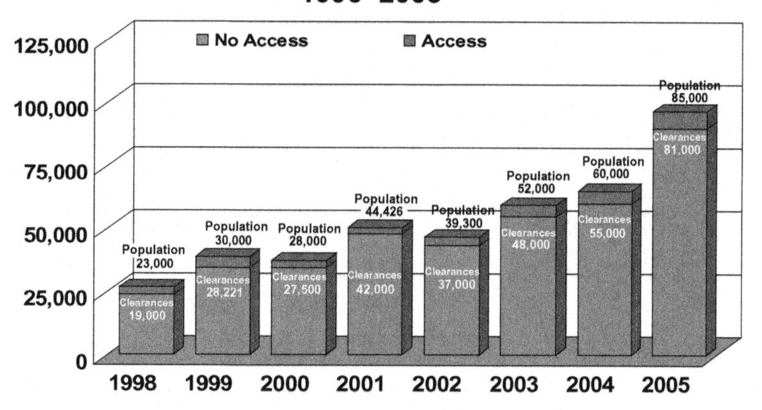

Sensitive Accesses 1998–2005

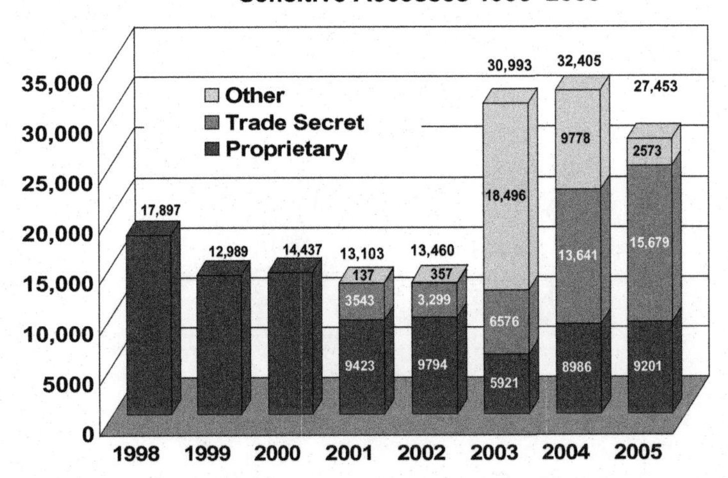

IWC Sensitive Material 2002–2005

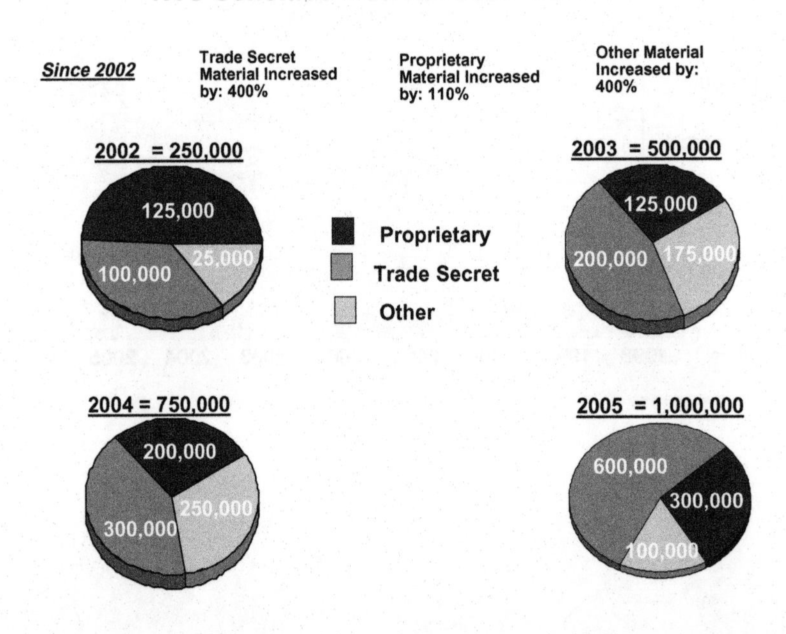

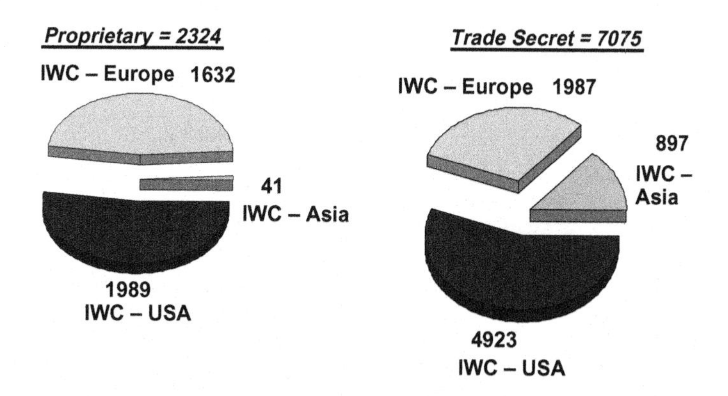
2005 Sensitive Areas Access by Region
Proprietary = 2324
IWC – Europe 1632
41
IWC – Asia
1989
IWC – USA
Trade Secret = 7075
IWC – Europe 1987
897
IWC – Asia
4923
IWC – USA

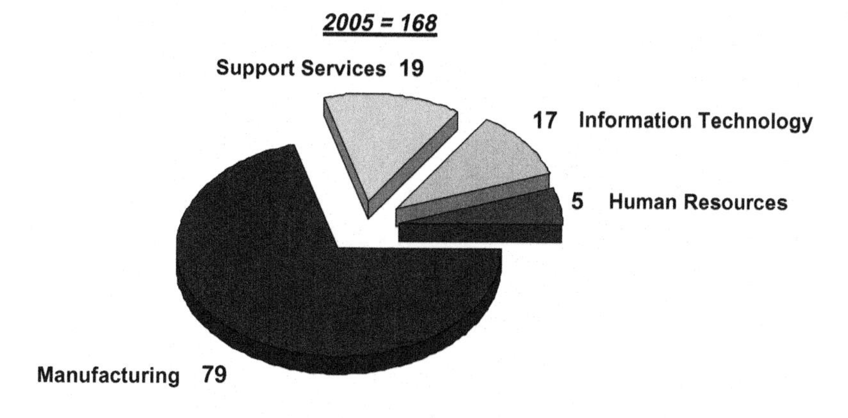
Total Foreign National Employees Working at IWC—USA
2005 = 168
Support Services 19
17 Information Technology
5 Human Resources
Manufacturing 79

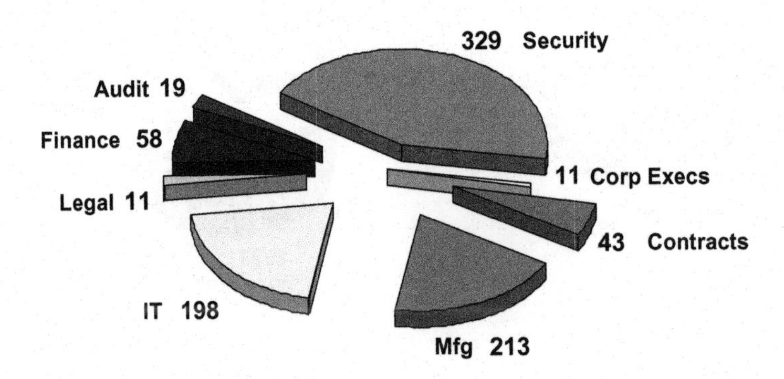
US Persons Working at IWC—US Corporate Headquarters
2005 = 769
329 Security
Audit 19
Finance 58
Legal 11
11 Corp Execs
43 Contracts
IT 198
Mfg 213

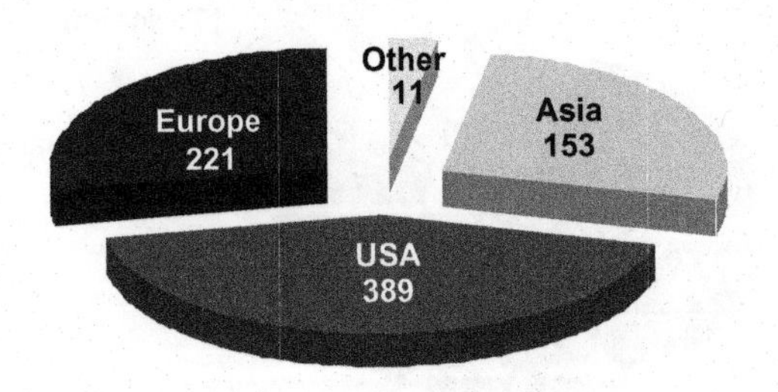
Expatriates & Dependents by Site
2005 = 593
Other 11
Europe 221
Asia 153
USA 389

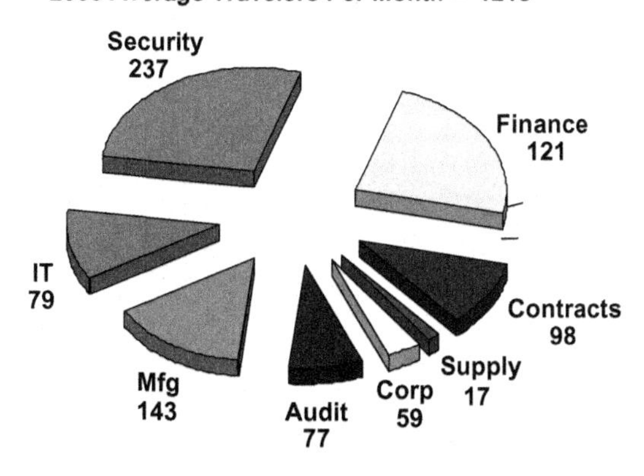
International Travelers
2005 Average Travelers Per Month = 1218
Security
237
Finance
121
IT
79
Contracts
98
Supply
17
Mfg
143
Audit
77
Corp
59

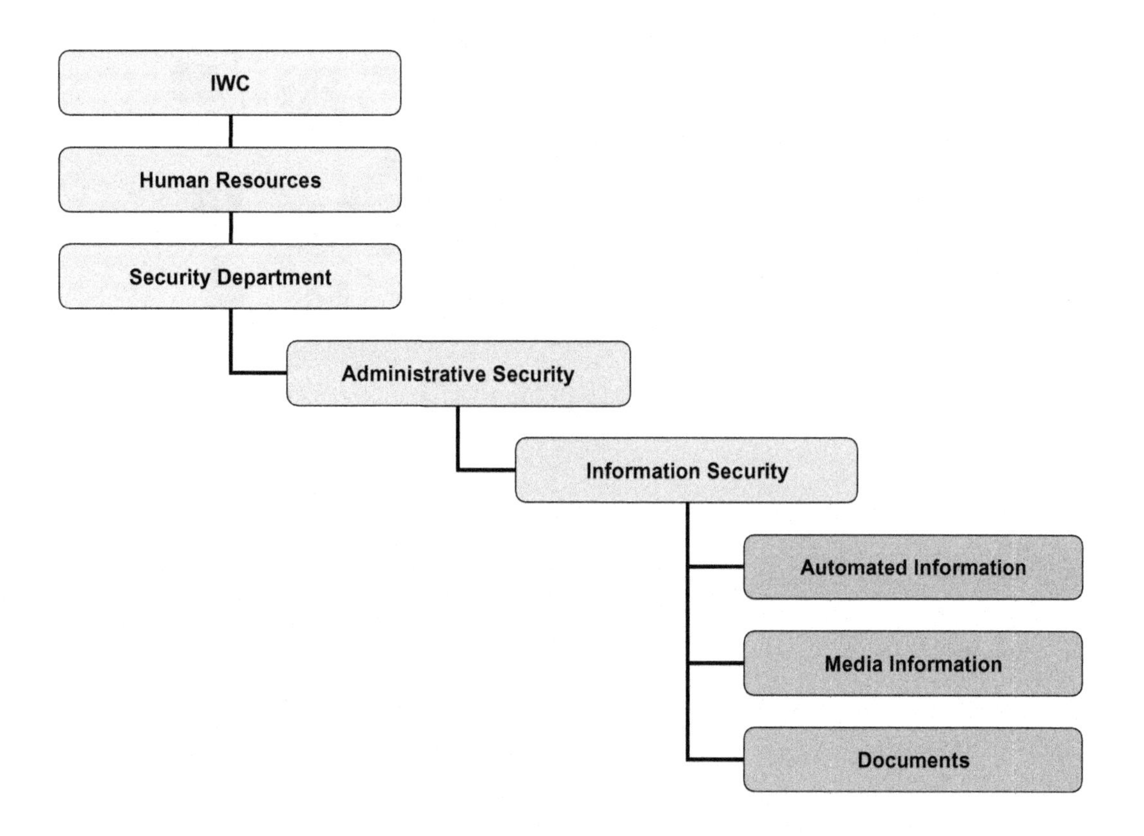
IWC
Human Resources
Security Department
Administrative Security
Information Security
Automated Information
Media Information
Documents

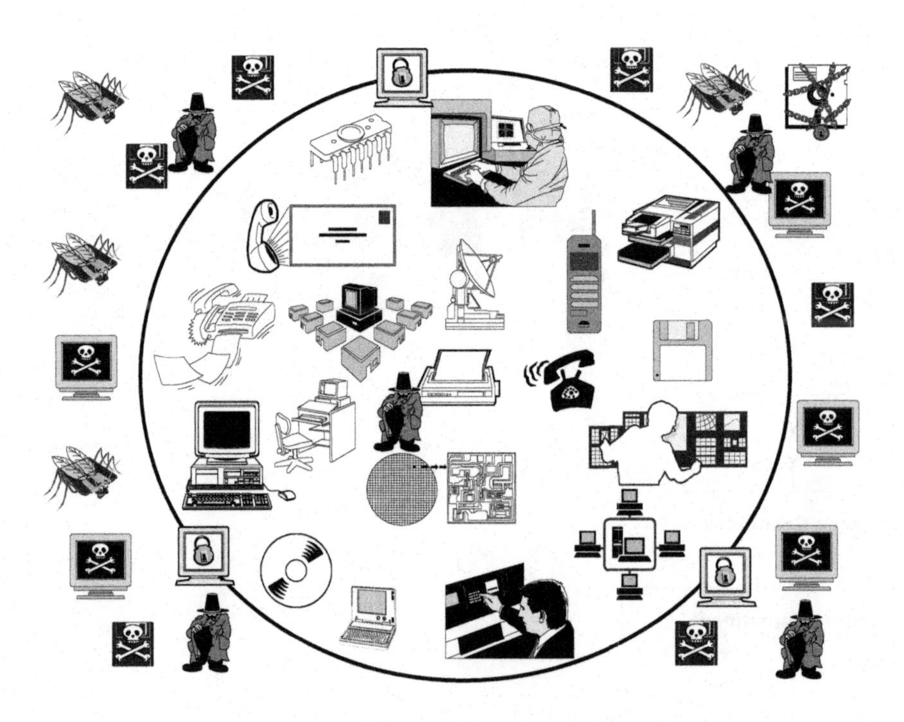

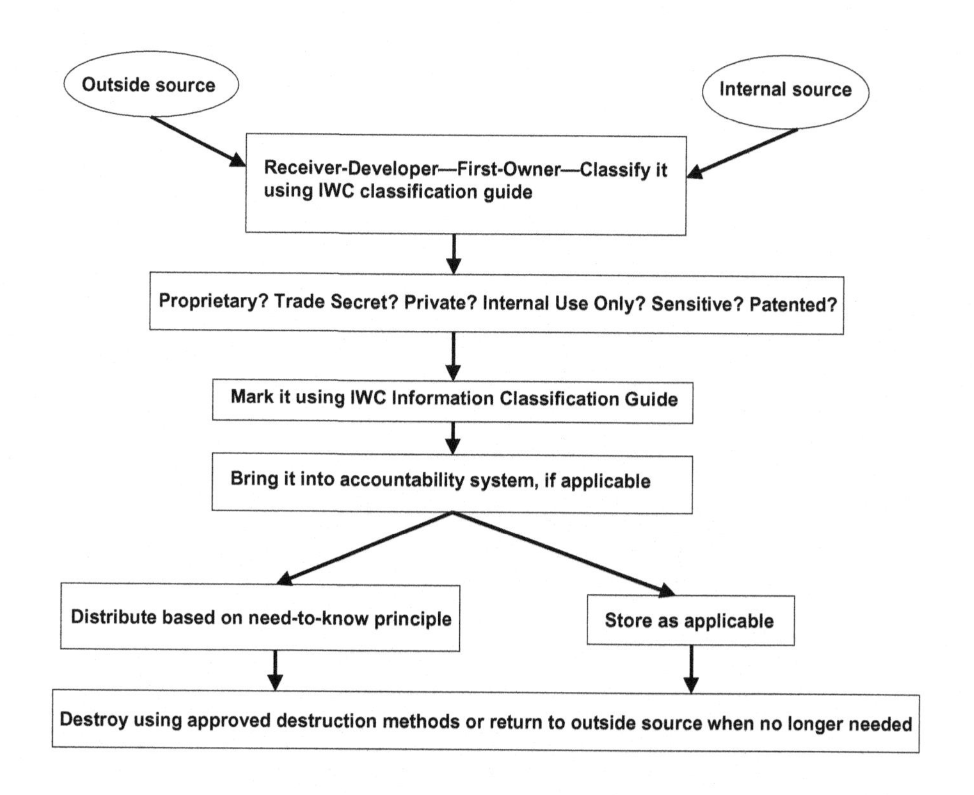

Number of IWC Proprietary Documents and The Cost of Storing Them—1st Quarter 2005

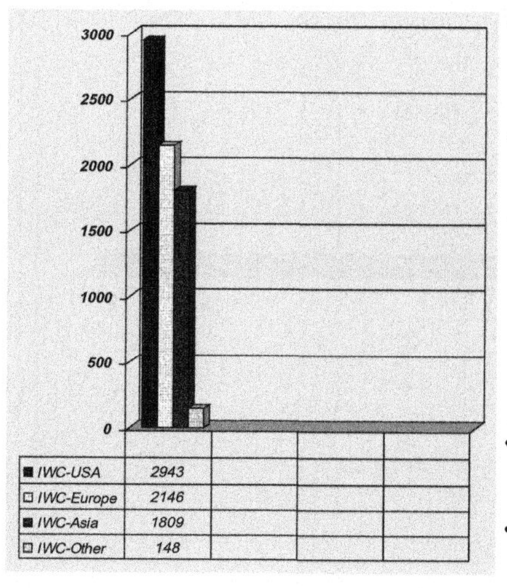

- There are a total of 7046 proprietary documents and other media, i.e., CDs, being stored throughout IWC facilities worldwide.
- The estimated cost of storing each information media per month, at all locations, includes labor, secure containers, and excluding processing time to take it in and out of storage, e.g., unlock the container, sign the container open-close sheet as open, lock the container, and sign it as closed, and other information assets handling procedures.
- The cost is approximately $15 per month per document or $45 per quarter.
- Total cost per month at all locations is approximately $105,690 and $1,268,280 per year.

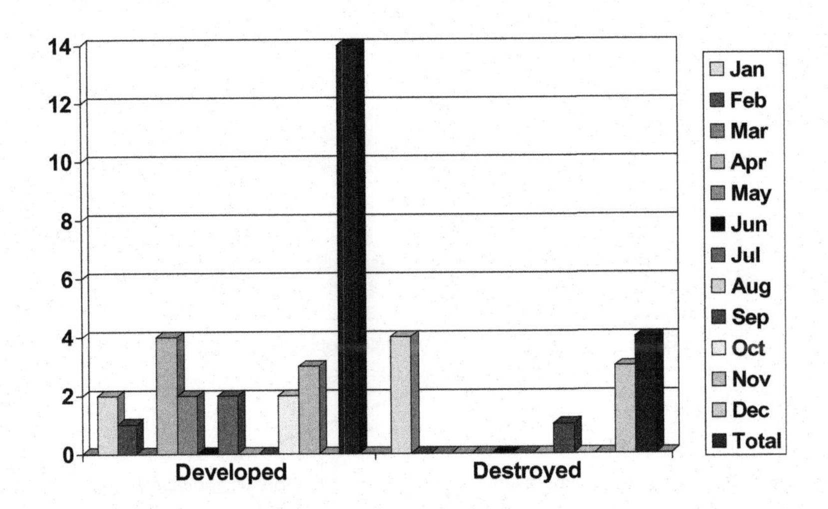
IWC Corporate Office Trade Secret Documents
Developed and Destroyed
14
12
10
8
6
4
2
0
Developed
Destroyed
Jan
Feb
Mar
Apr
May
Jun
Jul
Aug
Sep
Oct
Nov
Dec
Total

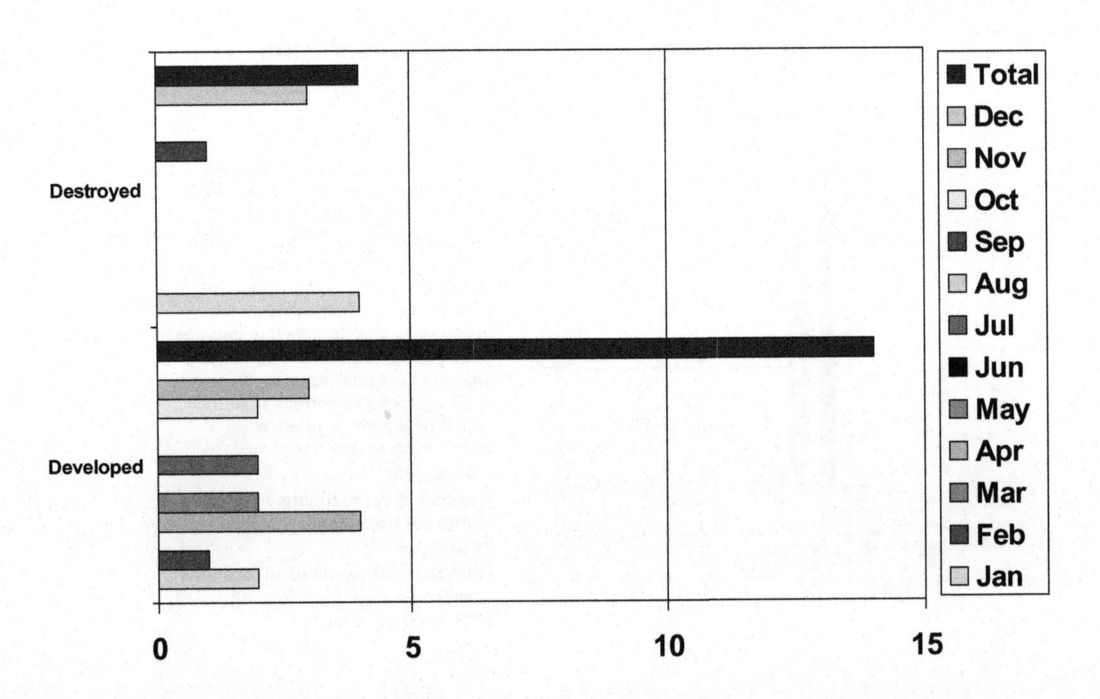
IWC Corporate Office Trade Secret Documents
Developed and Destroyed
Destroyed
Developed
0
5
10
15
Total
Dec
Nov
Oct
Sep
Aug
Jul
Jun
May
Apr
Mar
Feb
Jan

IWC Corporate Office Trade Secret Documents Developed and Destroyed

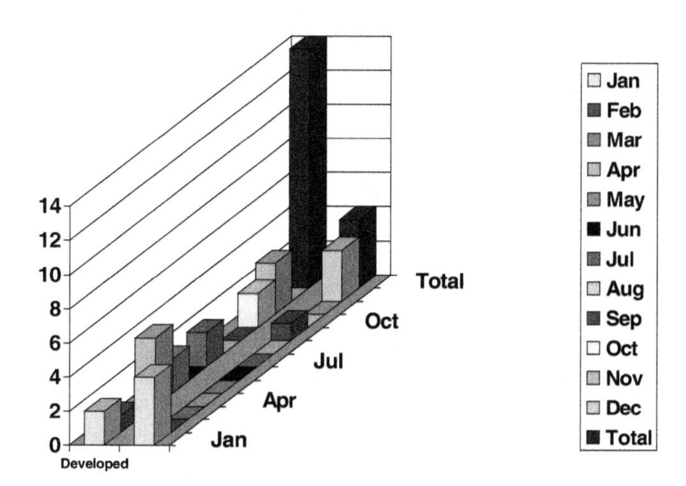

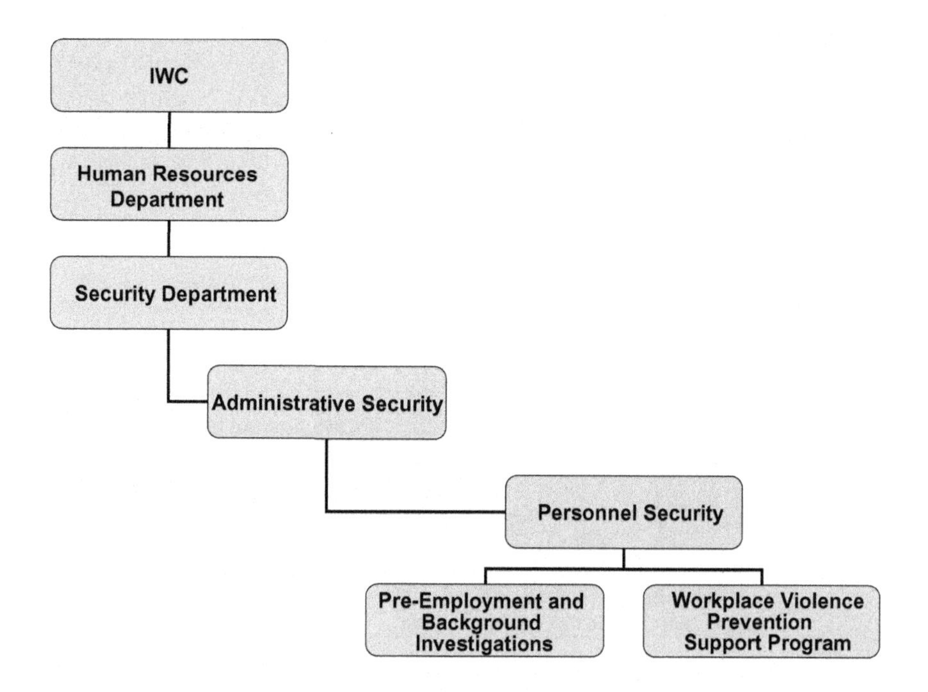

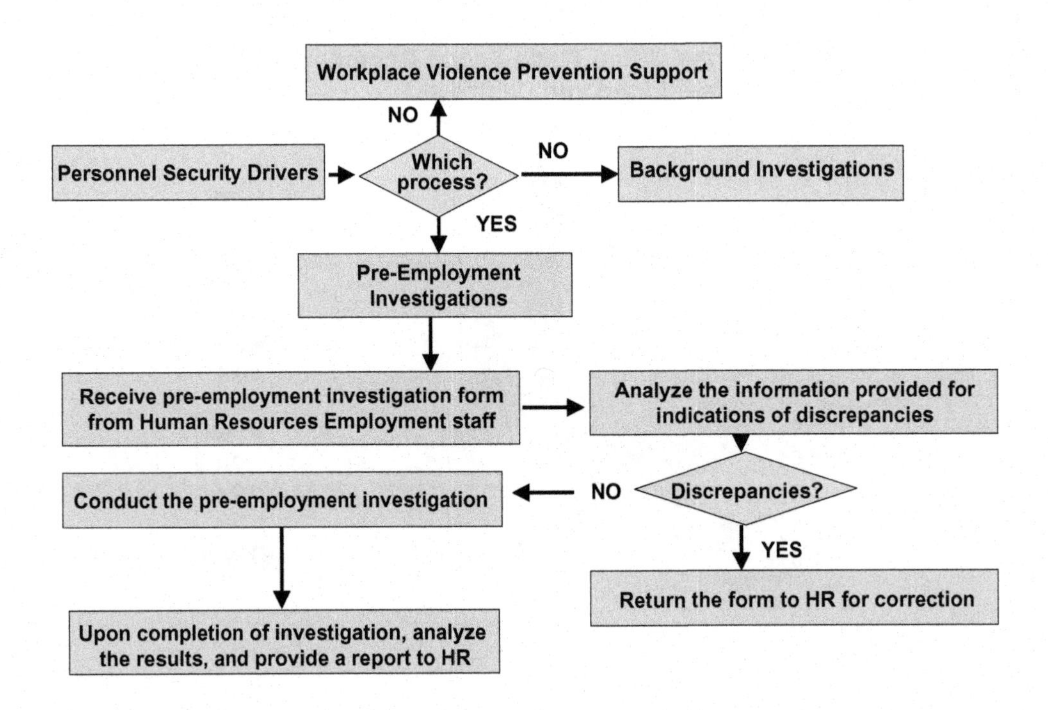
Workplace Violence Prevention Support
NO
Personnel Security Drivers
Which process?
NO
Background Investigations
YES
Pre-Employment Investigations
Receive pre-employment investigation form from Human Resources Employment staff
Analyze the information provided for indications of discrepancies
Conduct the pre-employment investigation
NO
Discrepancies?
YES
Return the form to HR for correction
Upon completion of investigation, analyze the results, and provide a report to HR

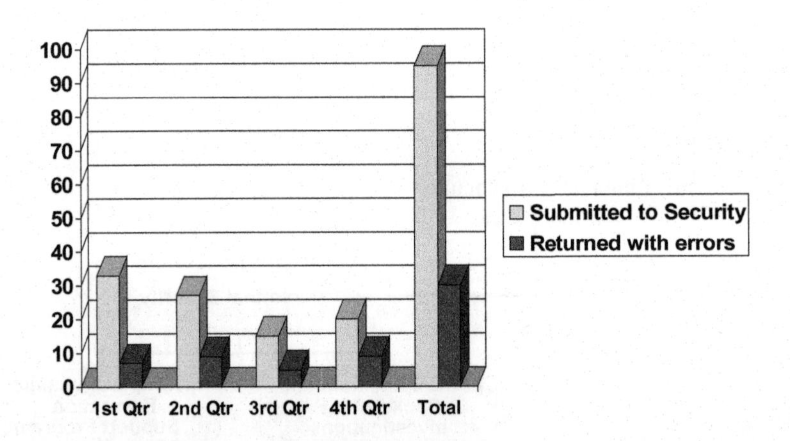
Pre-Employment Screening Forms—Quality Check
100
90
80
70
60
50
40
30
20
10
0
1st Qtr
2nd Qtr
3rd Qtr
4th Qtr
Total
Submitted to Security
Returned with errors

PRE-EMPLOYMENT METRICS DATA

Date Received	Name of HR sender	Name of applicant	Department	Returned to HR-Errors		Identify Error	Date Inves. Started	Date Inves. Completed	Derogatory Information		Returned to HR
				Yes	No				Yes	No	

Pre-Employment Investigations—2005
Average Completion Time per Case

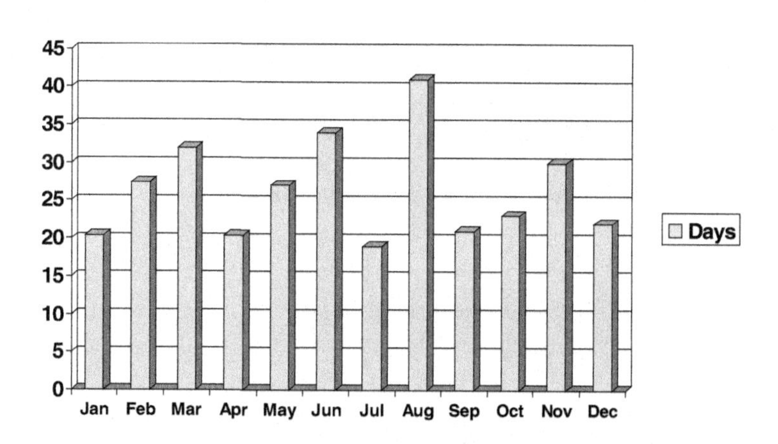

Pre-Employment Investigations—2005
Submitted—Negative Information

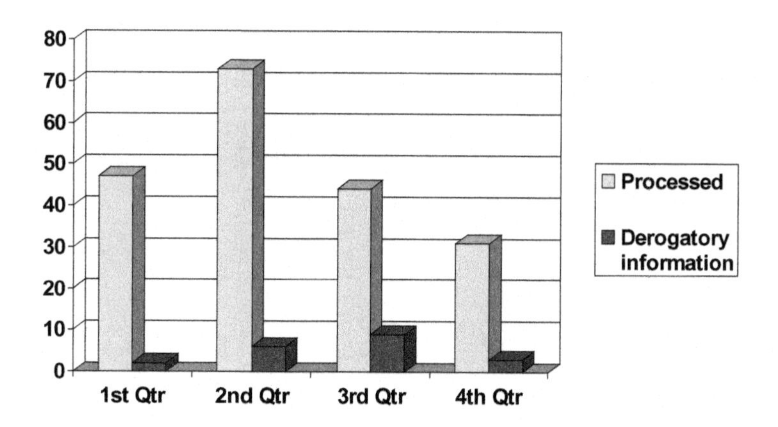

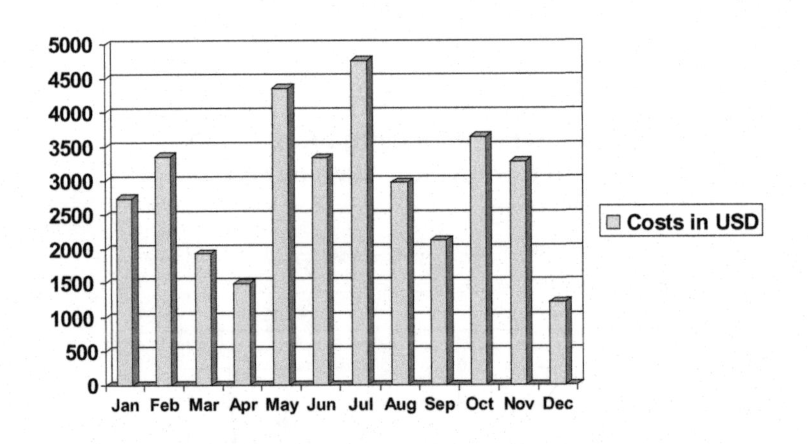
Pre-Employment Investigations—2005
Total Costs Per Month
5000
4500
4000
3500
3000
2500
2000
1500
1000
500
0
Jan
Feb
Mar
Apr
May
Jun
Jul
Aug
Sep
Oct
Nov
Dec
Costs in USD

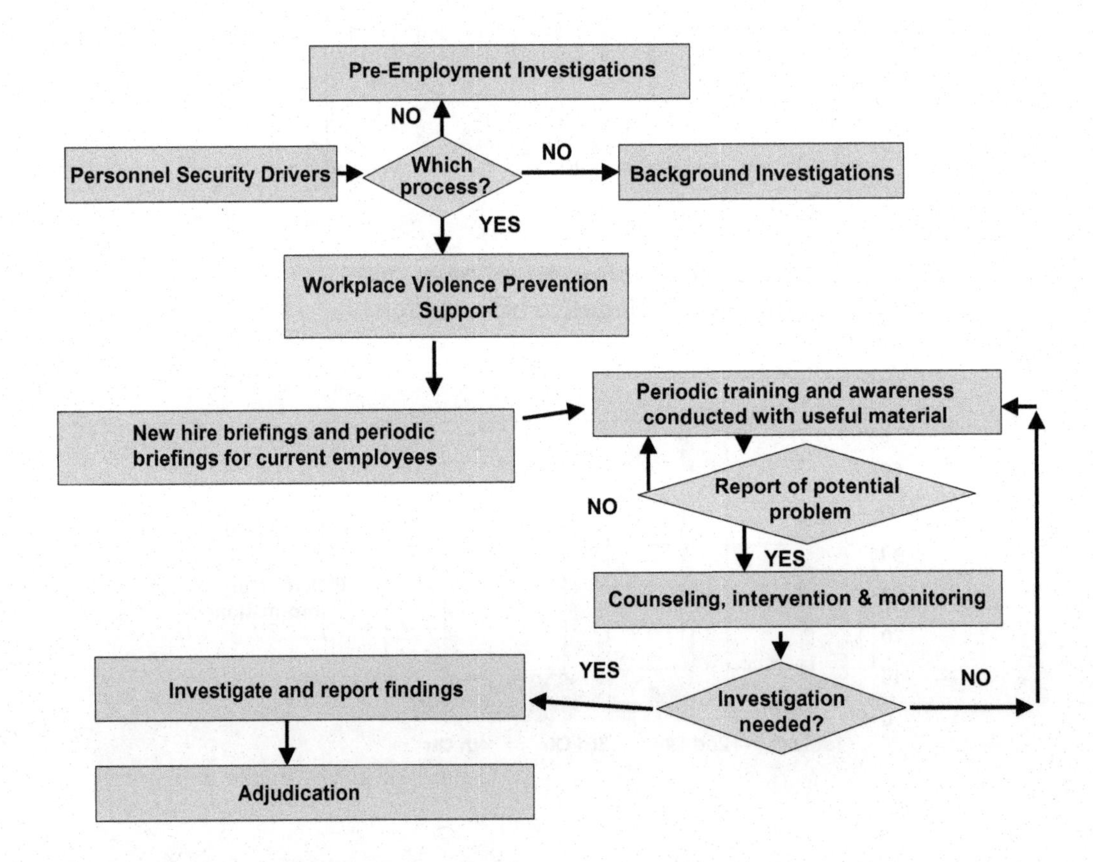
Pre-Employment Investigations
NO
Personnel Security Drivers
Which process?
NO
Background Investigations
YES
Workplace Violence Prevention Support
New hire briefings and periodic briefings for current employees
Periodic training and awareness conducted with useful material
Report of potential problem
NO
YES
Counseling, intervention & monitoring
YES
NO
Investigate and report findings
Investigation needed?
Adjudication

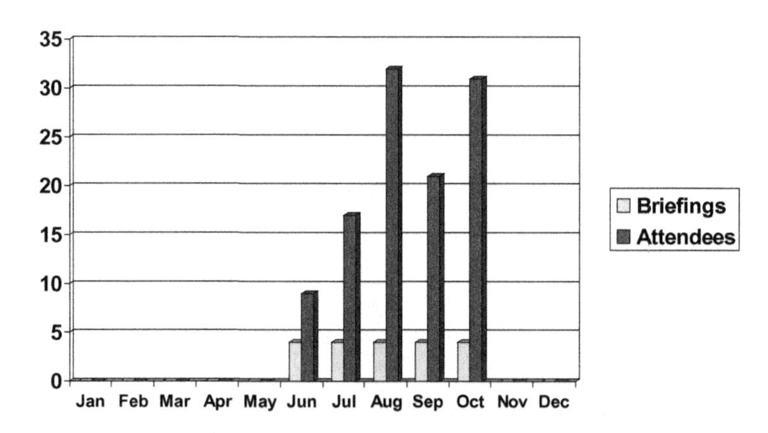
WPVPP Briefings and Attendees—2005
35
30
25
20
15
10
5
0
Jan Feb Mar Apr May Jun Jul Aug Sep Oct Nov Dec
Briefings
Attendees

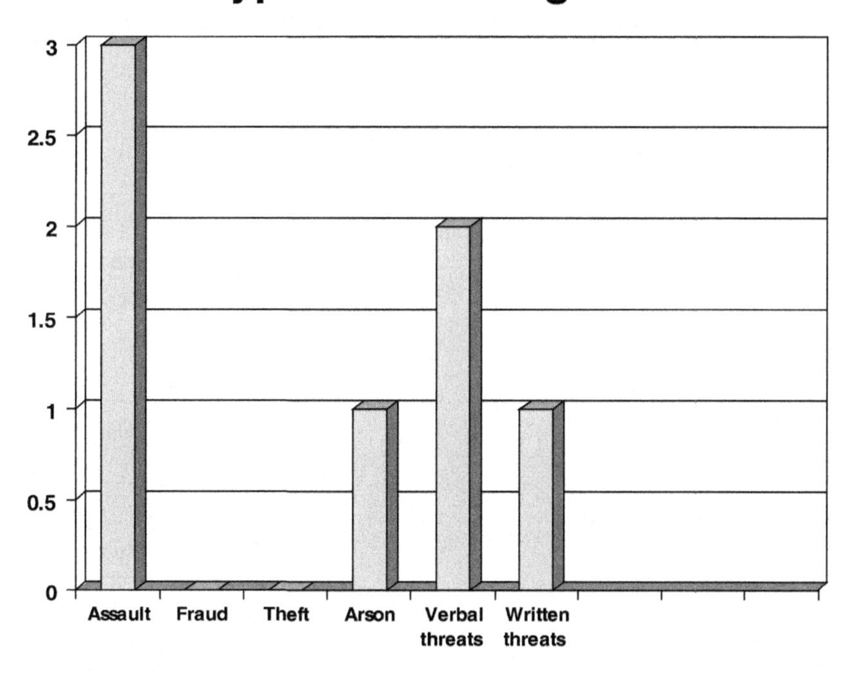
WPVPP Types of Investigations—2005
3
2.5
2
1.5
1
0.5
0
Assault
Fraud
Theft
Arson
Verbal threats
Written threats

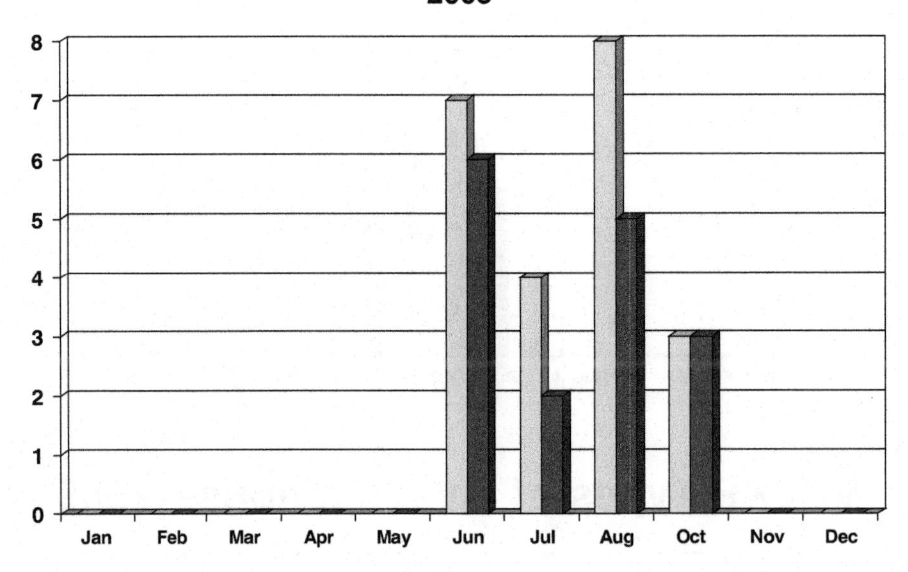
WPVPP—Complaints Received—Valid Complaints
2005
8
7
6
5
4
3
2
1
0
Jan
Feb
Mar
Apr
May
Jun
Jul
Aug
Oct
Nov
Dec

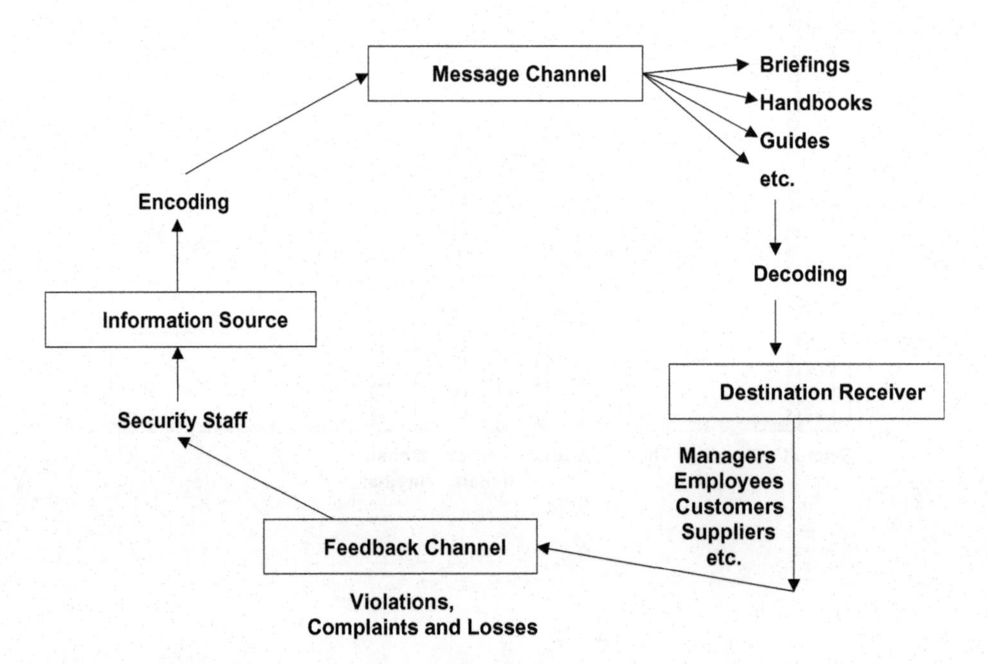
Communications Process
Message Channel
Briefings
Handbooks
Guides
etc.
Decoding
Destination Receiver
Managers
Employees
Customers
Suppliers
etc.
Feedback Channel
Violations,
Complaints and Losses
Security Staff
Information Source
Encoding

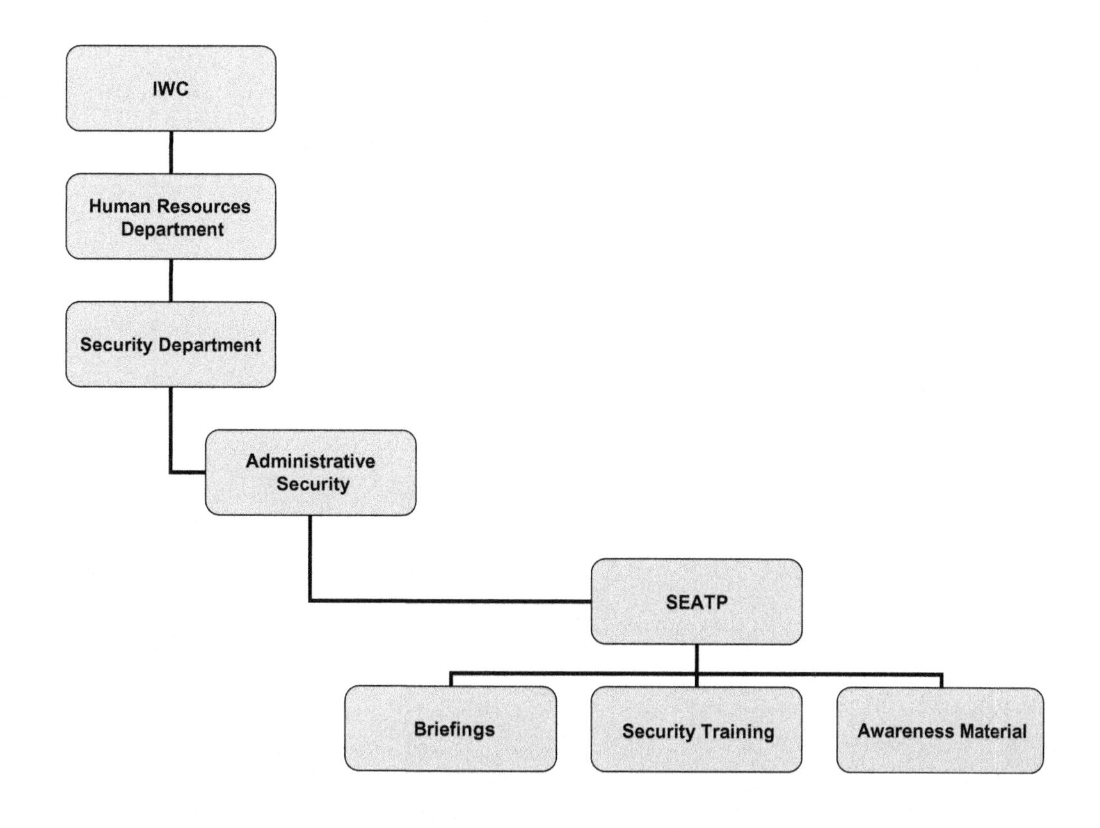

Security Education and Awareness Flow Chart

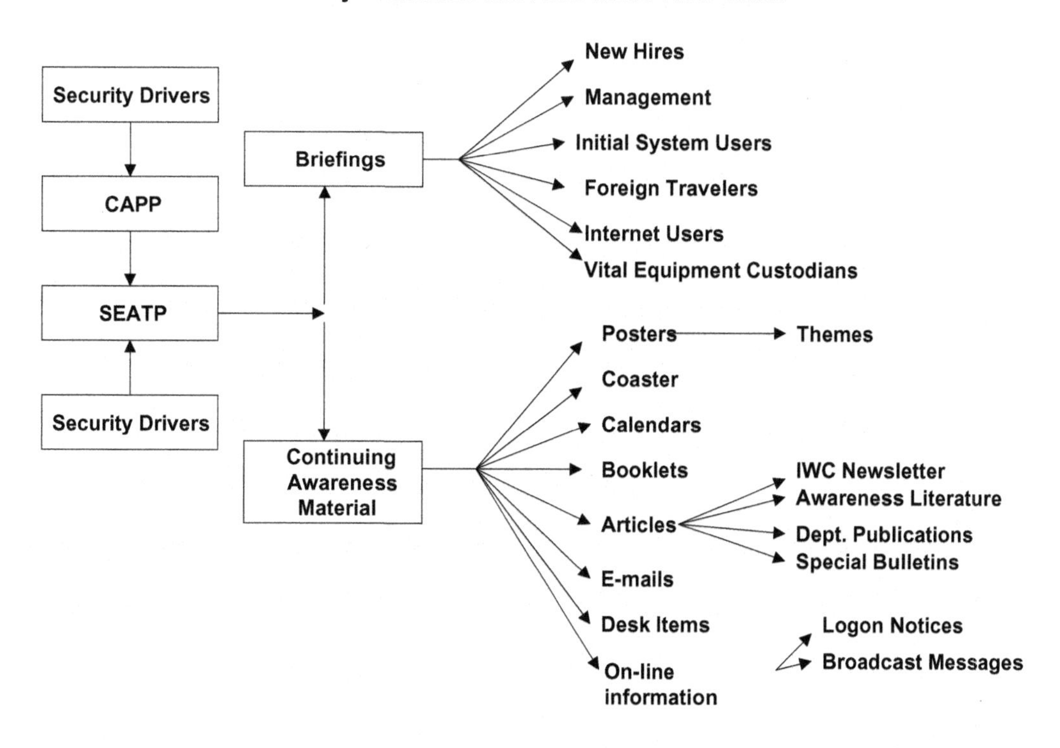

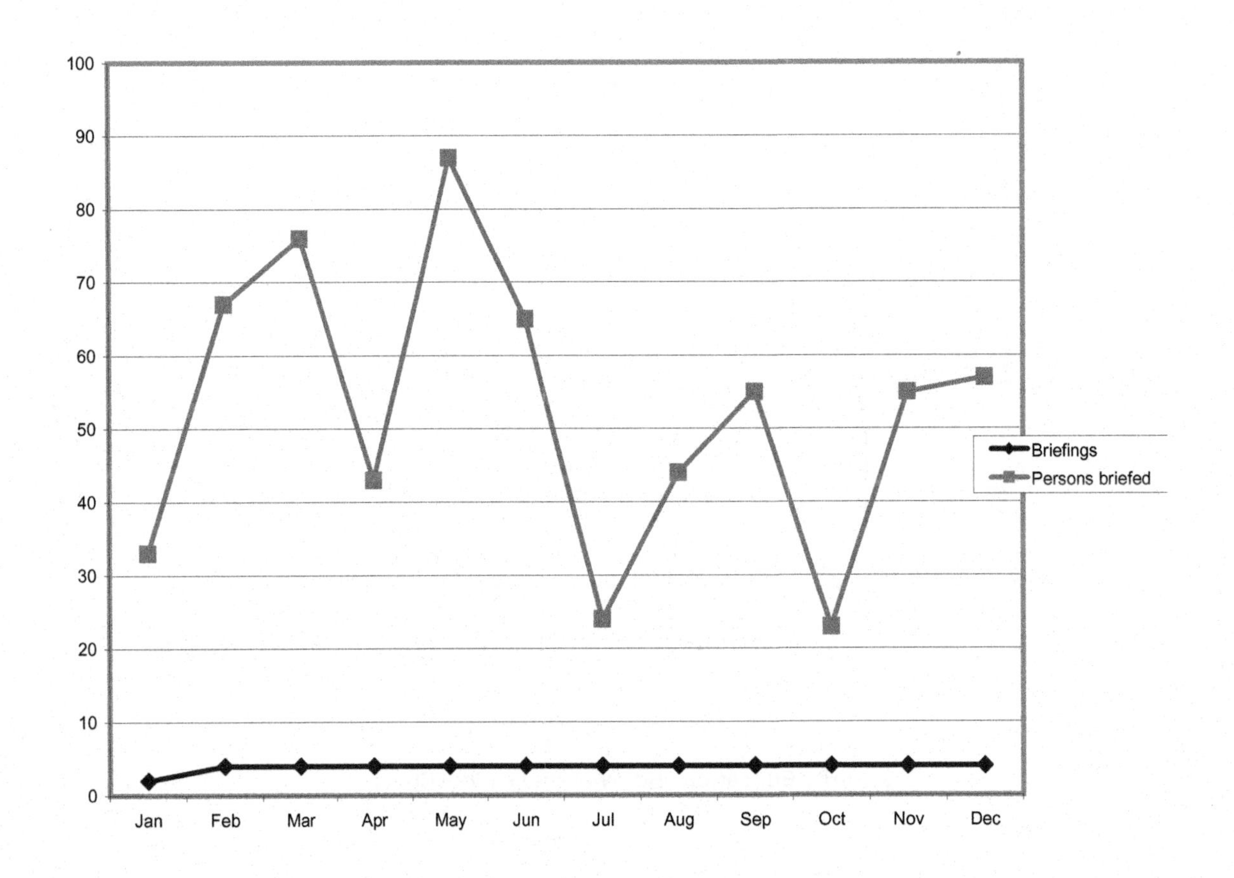

FW&A Annual Briefing Program

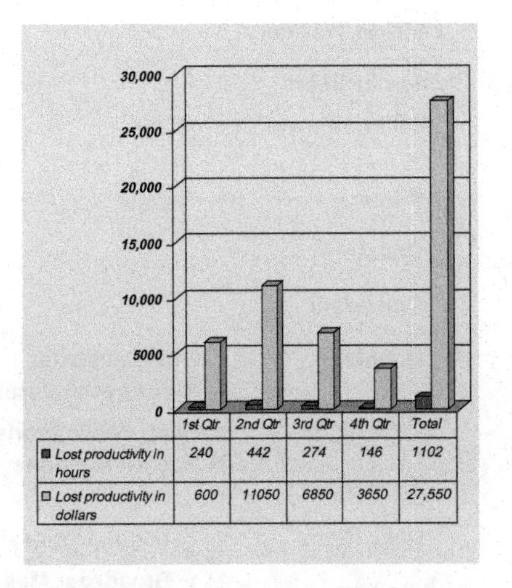

	1st Qtr	2nd Qtr	3rd Qtr	4th Qtr	Total
Lost productivity in hours	240	442	274	146	1102
Lost productivity in dollars	600	11050	6850	3650	27,550

- The FW&A briefing project was launched per CEO direction based on new federal law requiring FW&A briefings to all employees within 90 days and annually thereafter. Proof of compliance must be available.
- During the first year of annual briefings all 551 IWC corporate employees were briefed. Similar briefings were held at all IWC US locations.
- Total cost in lost productivity to IWC Corporate Office was $44,300 for 1 year.
- Productivity lost based on 1 h briefing, 1 h round trip to briefing site and back, and average IWC wage with benefits of $50.00 per hour.

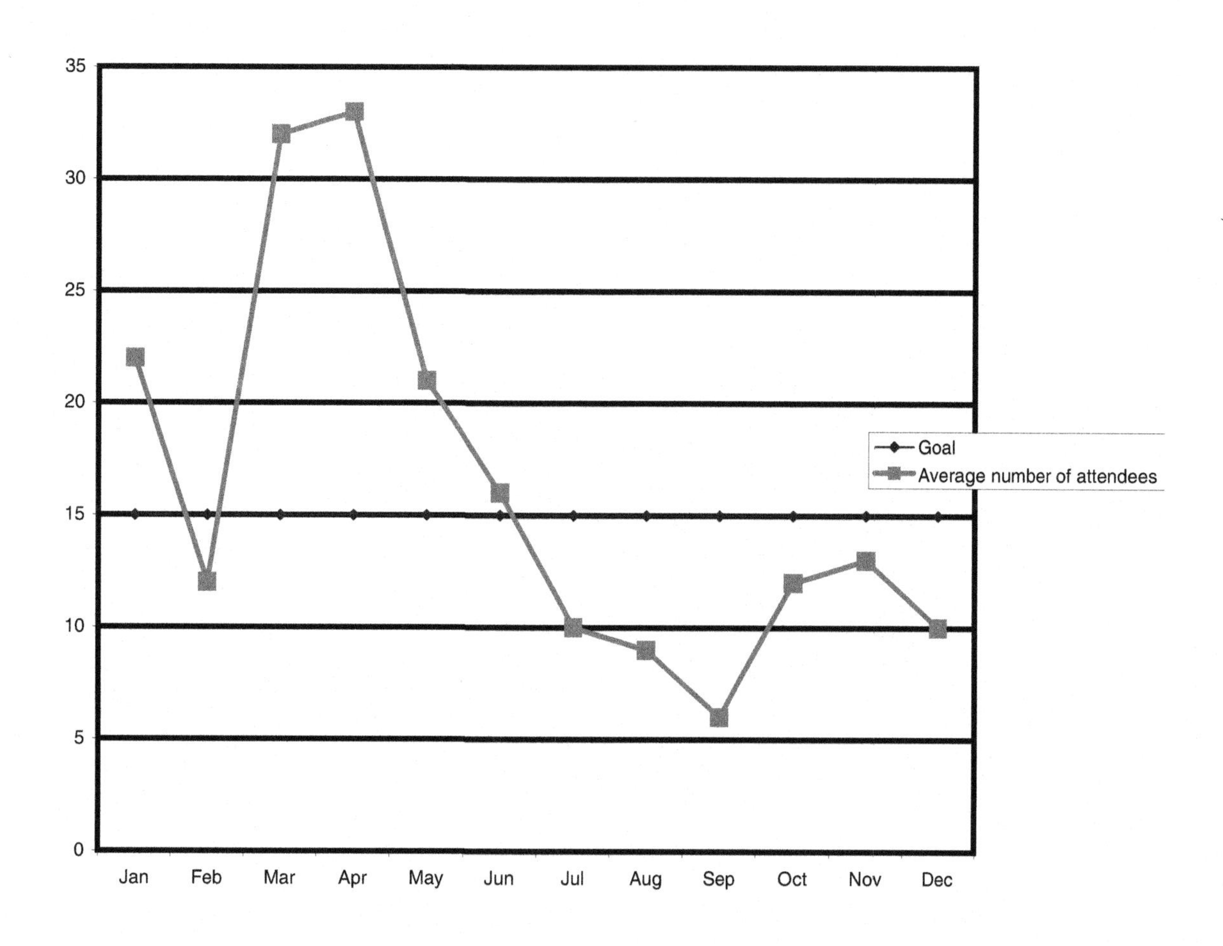

High-Value Assets Losses—2004–2005

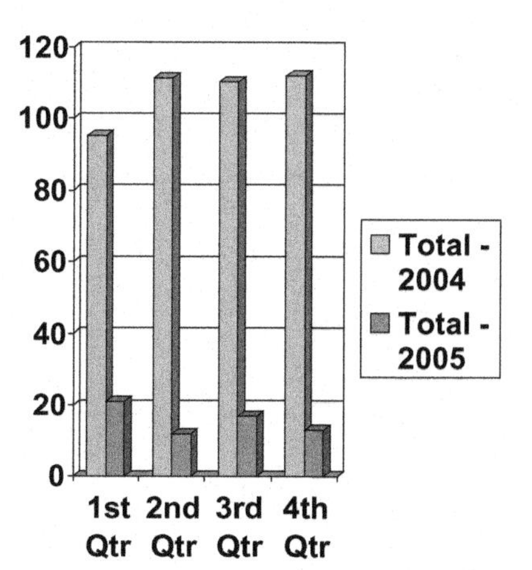

- High-value assets losses in 2004 include cell phones, PDAs, and notebook computers containing sensitive IWC information. None were recovered.
- In 2005, change in CAPP policy requiring employees' pay for items if negligence shown.
- Awareness briefings and materials modified to emphasize need to protect these assets and consequences of not protecting them.
- A downward trend in losses occurred as the result of an investment in awareness material and briefings. This action contributed to an overall savings as the cost of past losses was greater than the cost of awareness briefings.

High-Technology Assets Thefts and Losses

Date	Employee	Department	Item	Lost When	Where	Why	How	Class of info.	Price of item

High-Technology Assets Losses—2004–2005

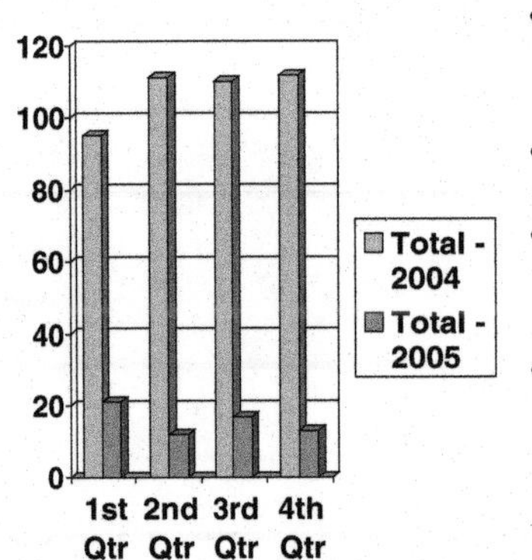

- High-technology assets losses in 2004 include cell phones, PDAs, and notebook computers containing sensitive IWC information. None were recovered.
- In 2005, change in CAPP policy requiring employees' pay for items if negligence shown.
- Awareness briefings and materials modified to emphasize need to protect these assets and consequences of not protecting them.
- A downward trend in losses resulted in a cost of awareness material of $327 and briefing change at a cost of $43 for $370 which contributed to a net saving of $37,872

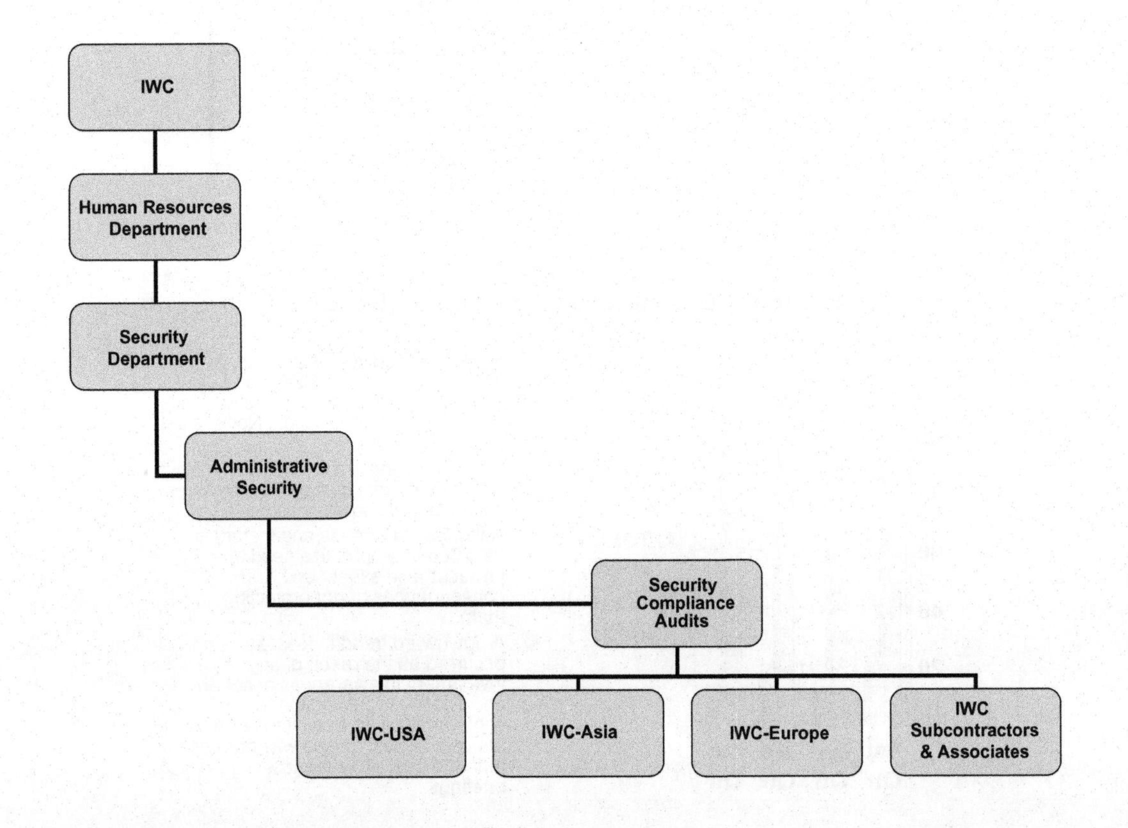

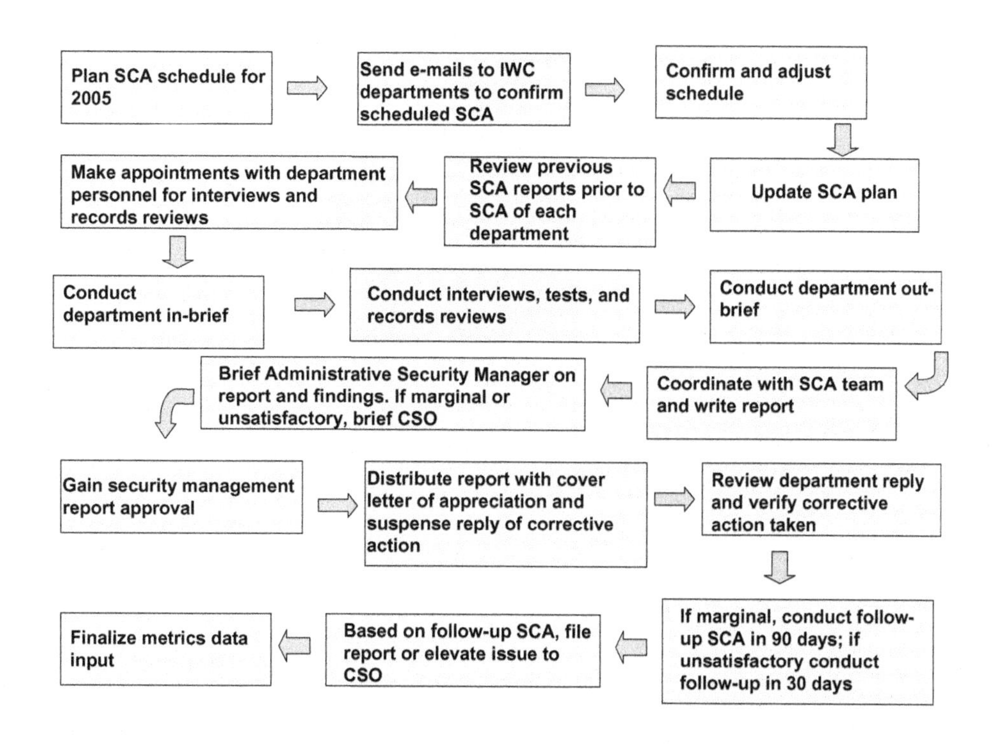
Plan SCA schedule for 2005
Send e-mails to IWC departments to confirm scheduled SCA
Confirm and adjust schedule
Update SCA plan
Review previous SCA reports prior to SCA of each department
Make appointments with department personnel for interviews and records reviews
Conduct department in-brief
Conduct interviews, tests, and records reviews
Conduct department out-brief
Coordinate with SCA team and write report
Brief Administrative Security Manager on report and findings. If marginal or unsatisfactory, brief CSO
Gain security management report approval
Distribute report with cover letter of appreciation and suspense reply of corrective action
Review department reply and verify corrective action taken
If marginal, conduct follow-up SCA in 90 days; if unsatisfactory conduct follow-up in 30 days
Based on follow-up SCA, file report or elevate issue to CSO
Finalize metrics data input

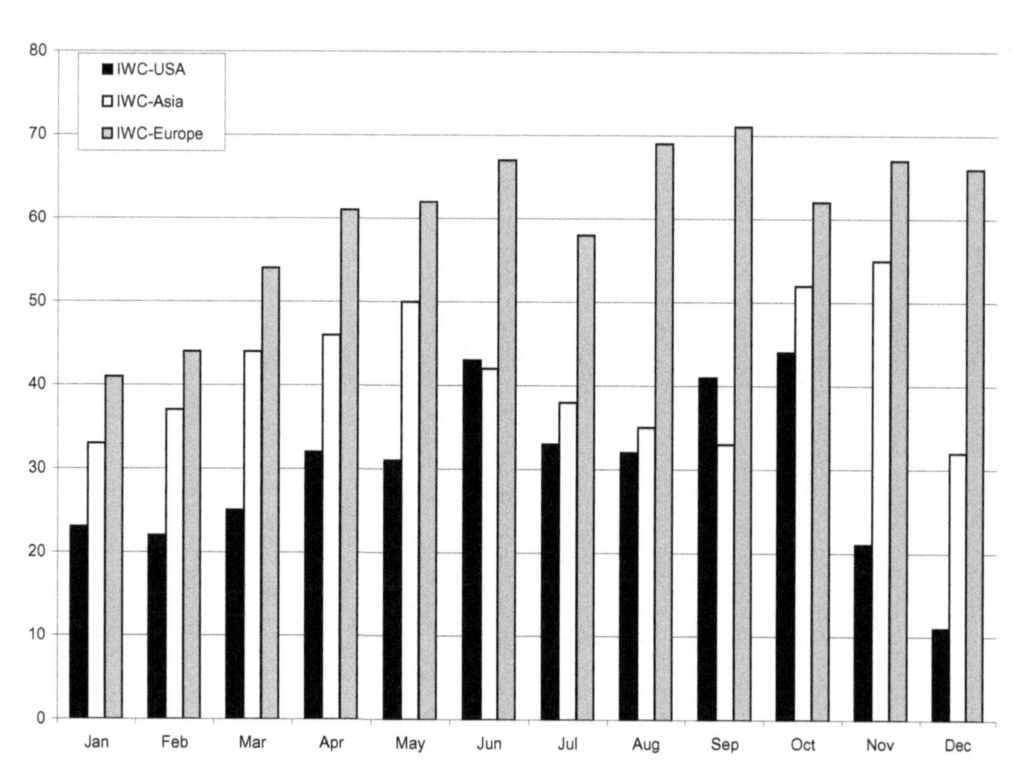
IWC-USA
IWC-Asia
IWC-Europe
0
10
20
30
40
50
60
70
80
Jan
Feb
Mar
Apr
May
Jun
Jul
Aug
Sep
Oct
Nov
Dec

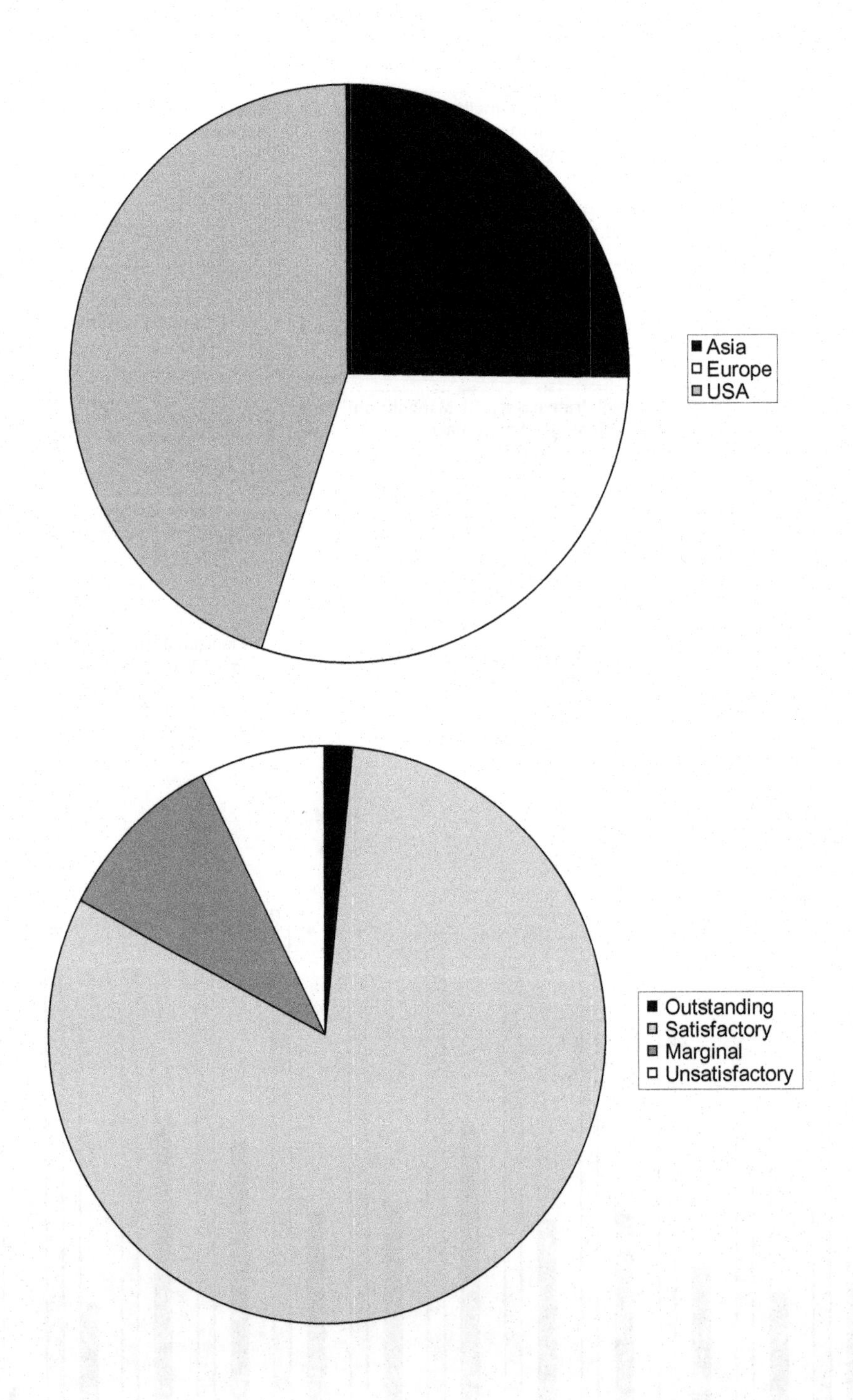
Asia
Europe
USA
Outstanding
Satisfactory
Marginal
Unsatisfactory

Security Compliance Audit Findings—2005

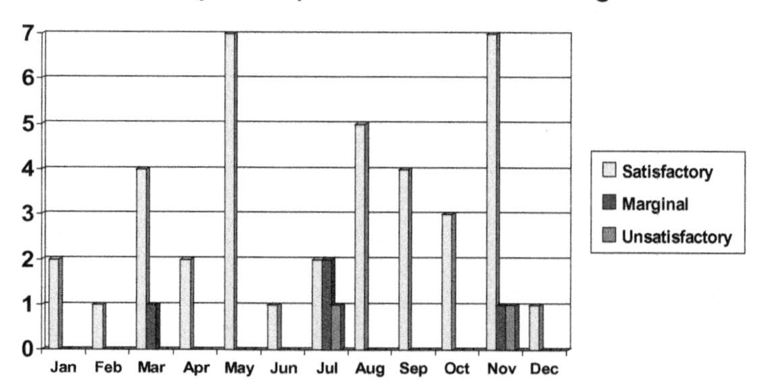

Security Compliance Audit Findings by Individual IWC Departments (1–9)

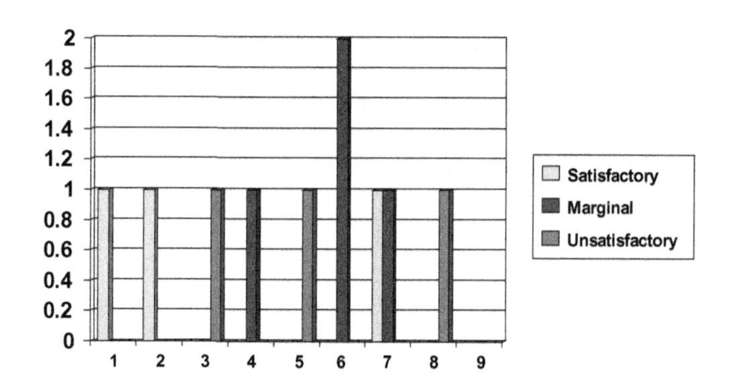

Security Compliance Audits— Hours Expended—per Department

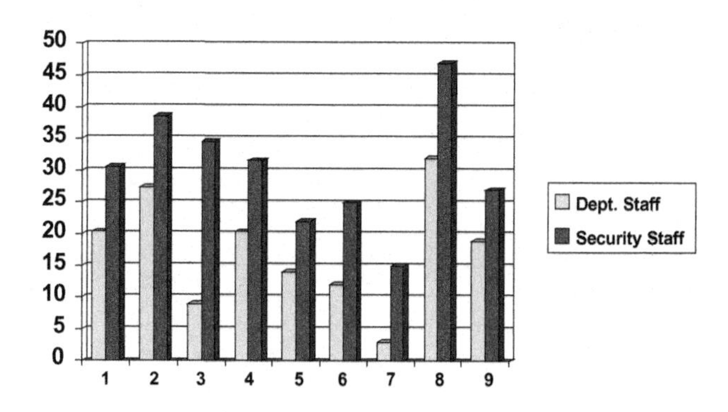

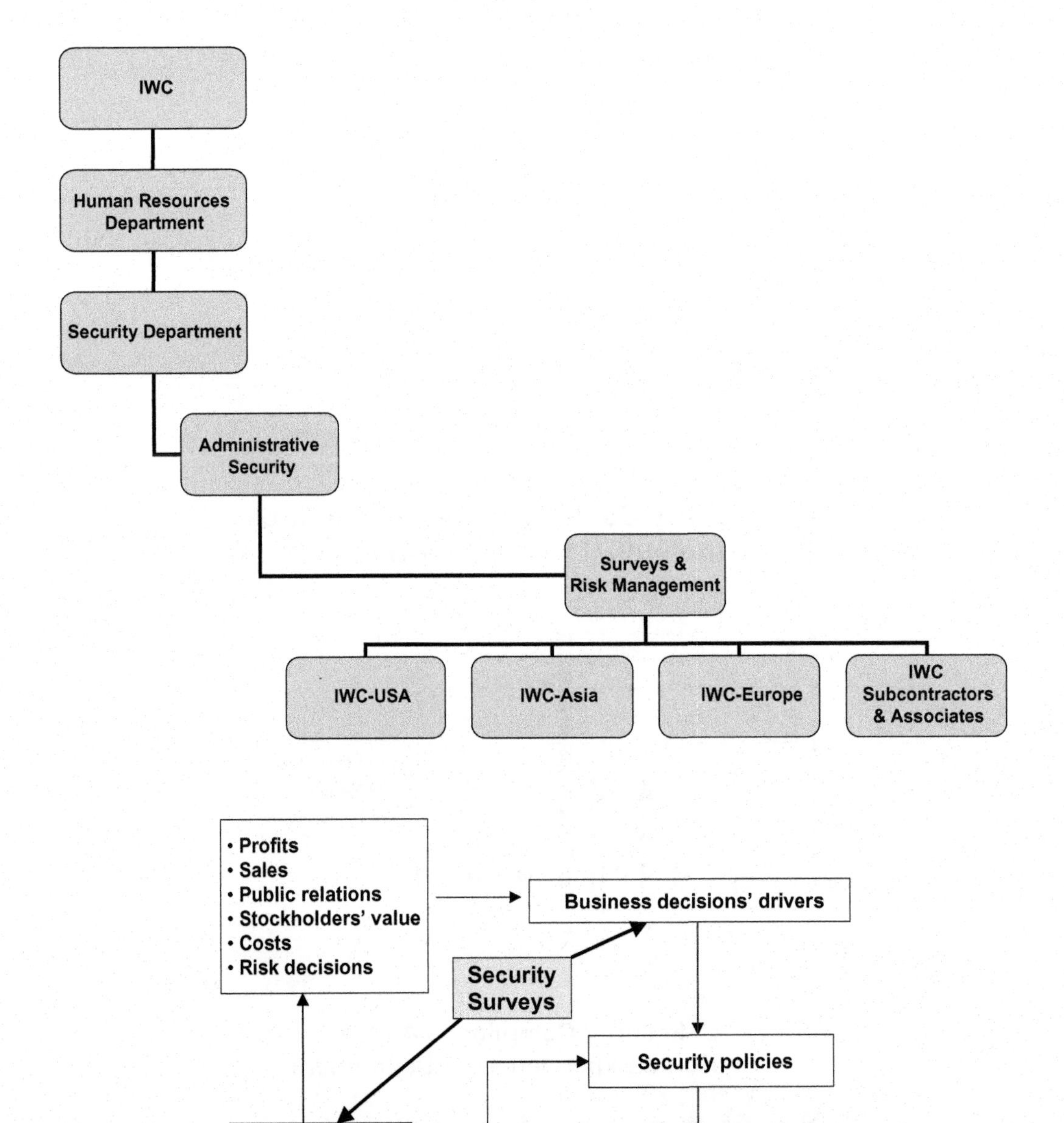

• Profits
• Sales
• Public relations
• Stockholders' value
• Costs
• Risk decisions

Business decisions' drivers

Security Surveys

Security policies

Risk management

Security procedures

Processes

Plans

Projects

• Risk assessment
• Risk analyses
• Best security practices

Primary driver: Potential threats to vulnerable assets leading to unknown risks
↓
Identification of assets group to be surveyed
↓
Operations plan developed
↓
Identify threat agents to the identified assets
↓
Conduct controlled attacks to the assets
↓
Identify the vulnerabilities of the assets and their associated protective mechanisms
↓
Identify the risks to those assets
↓
Identify the mitigation costs to reduce risks to acceptable level
↓
Identify and correct deficiencies that allowed for successful attacks
↓
Write report and brief management

The Number of High–Value Assets Crime Prevention Surveys Conducted—2005

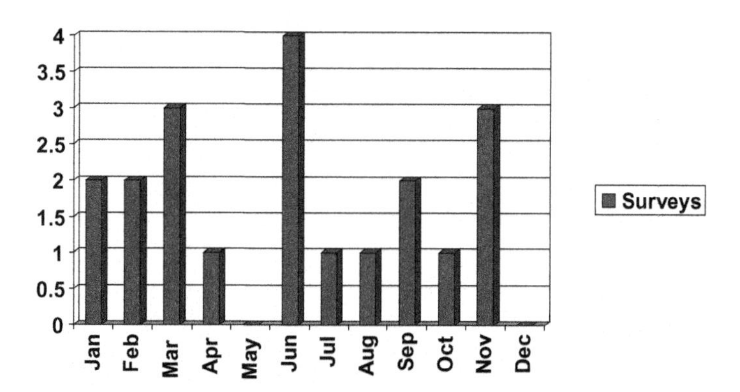

IWC Security Surveys by Region—2005

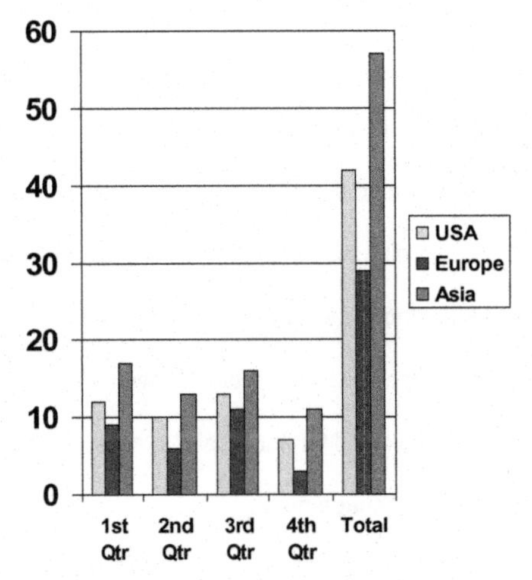

- There were a total of 128 security surveys conducted worldwide in 2005:
- USA – 41
- Europe – 30
- Asia – 57
- Cost of all surveys conducted was $134,743
- Risks to assets valued at $3,235,586
- Security surveys are part of value-added security services protecting assets using a proactive cost–benefits approach

Types of IWC Assets Surveyed—2005—by Region

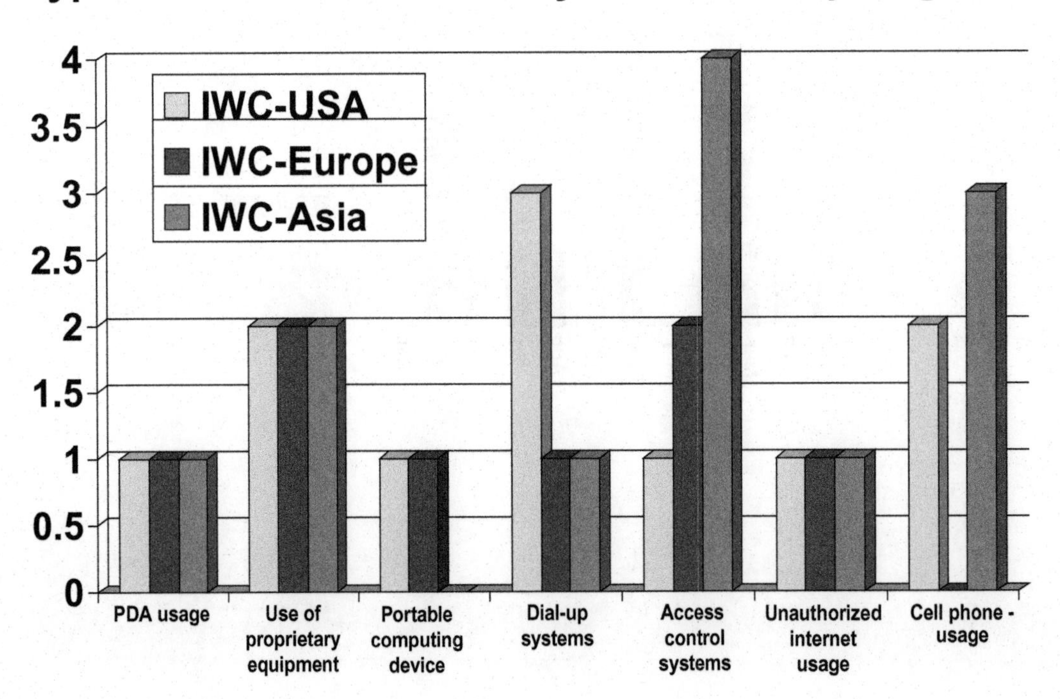

PDAs—Total Value—Total Lost/Stolen

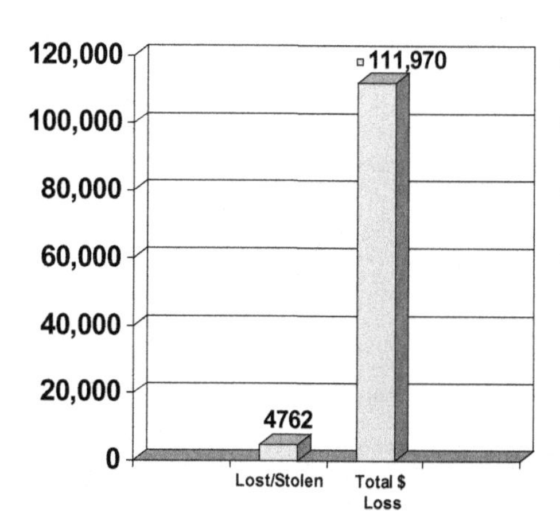

- 4762 PDAs lost–stolen: in all cases due to carelessness.
- No employee reimbursement requirement in place.
- Loss of $111,970 in purchase value of PDAs.
- Potential compromise of lost, sensitive information maintained on the PDAs is in the millions of dollars, e.g., IWC Customer's Lists, Executive Management home phone numbers.

2005 Costs of Security Surveys vs Savings to IWC

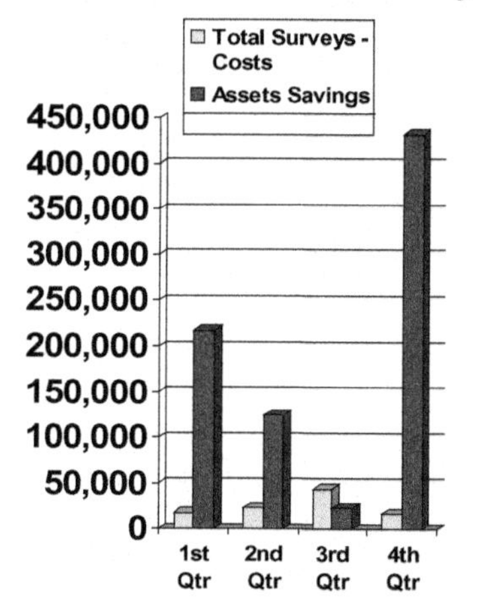

- The costs of the security surveys are actual costs of labor and resources for all regions by quarter.
- The total assets savings were calculated by identifying the assets' value that was demonstrated to be highly vulnerable to loss (destruction or theft).
- Proven highly vulnerable through use of known threat agents' techniques.
- Average ROI was $217,489 per survey.

Deficiencies in IWC Dial-up Systems

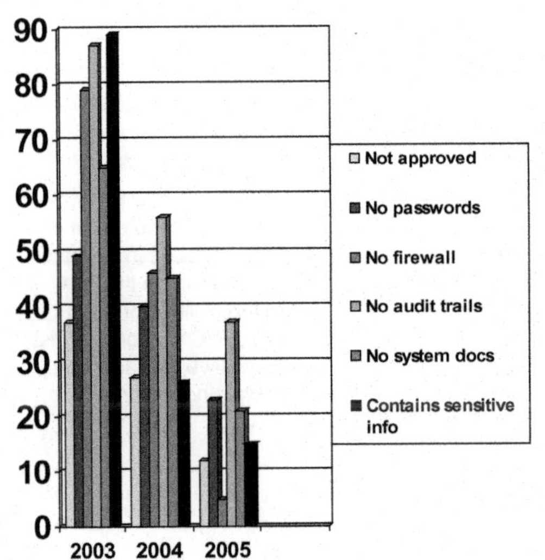

- A 2003 security survey found this is a major problem since annual surveys began
- The deficiencies noted included recurring problems
- Systems were vulnerable to external threat agents' attacks
- Numerous systems contained sensitive information that on at least two occasions were hacked and not reported

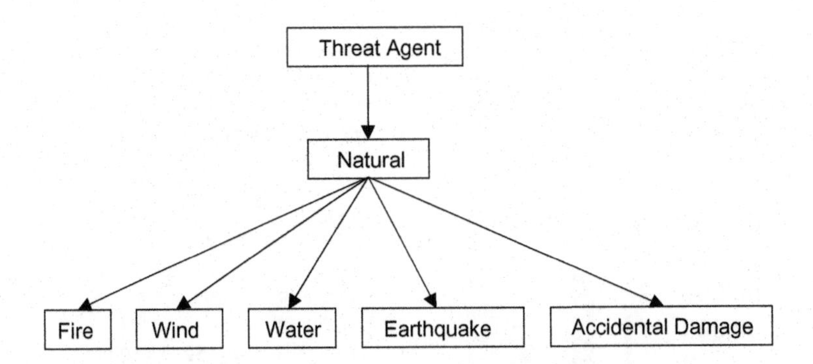

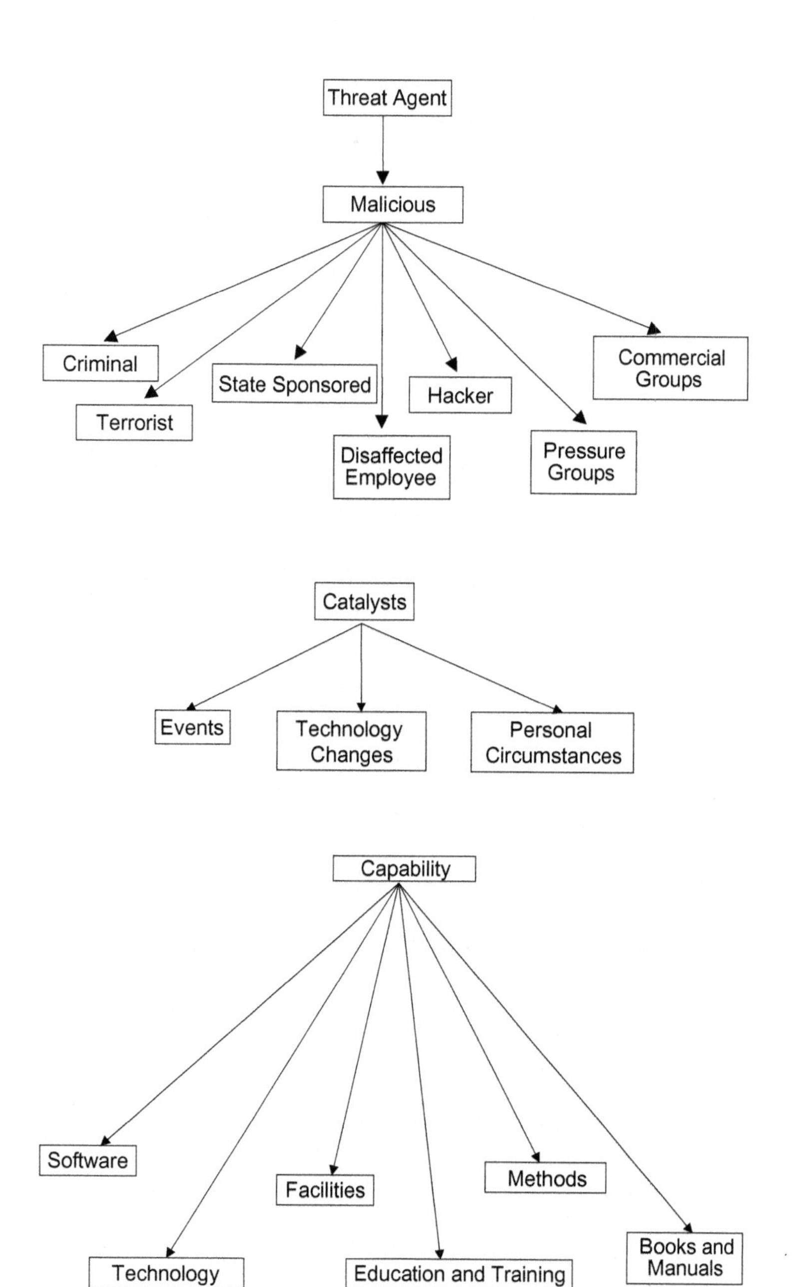
Threat Agent
Malicious
Criminal
Terrorist
State Sponsored
Disaffected Employee
Hacker
Pressure Groups
Commercial Groups
Catalysts
Events
Technology Changes
Personal Circumstances
Capability
Software
Technology
Facilities
Education and Training
Methods
Books and Manuals

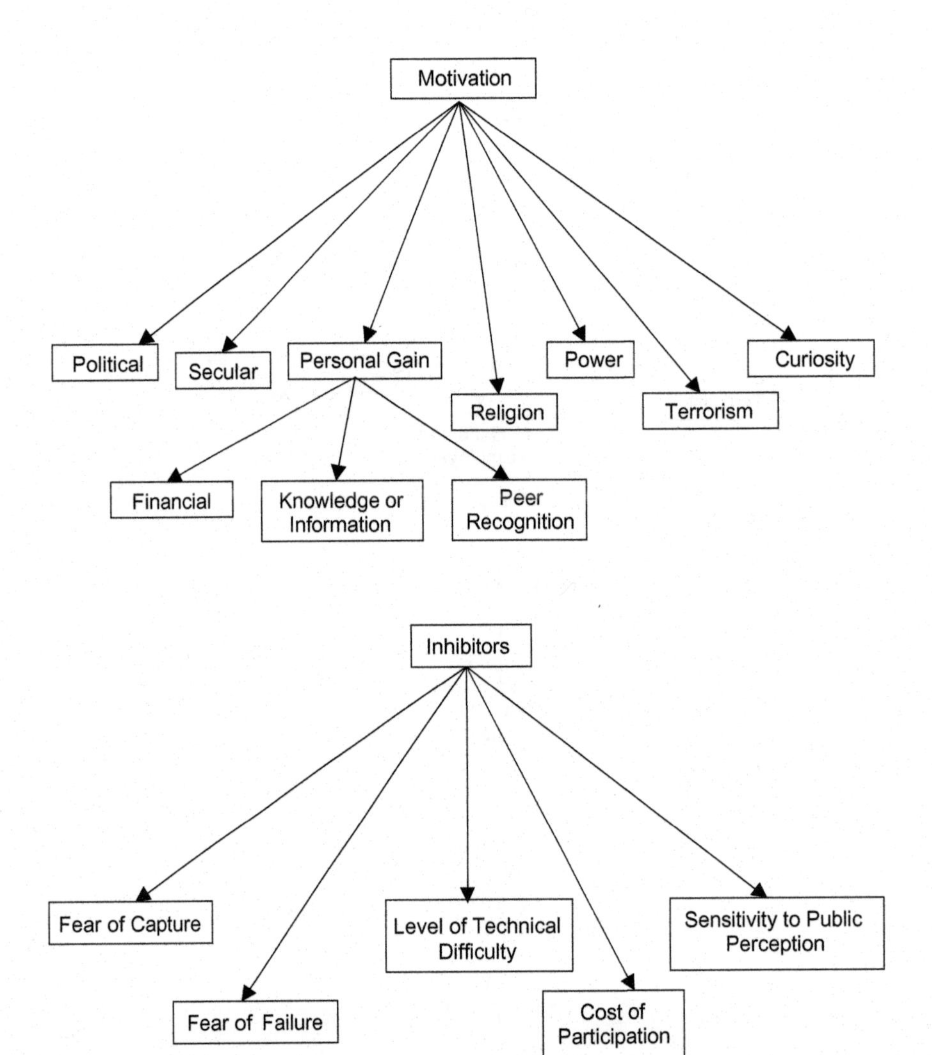
Motivation
Political
Secular
Personal Gain
Religion
Power
Terrorism
Curiosity
Financial
Knowledge or Information
Peer Recognition
Inhibitors
Fear of Capture
Fear of Failure
Level of Technical Difficulty
Cost of Participation
Sensitivity to Public Perception

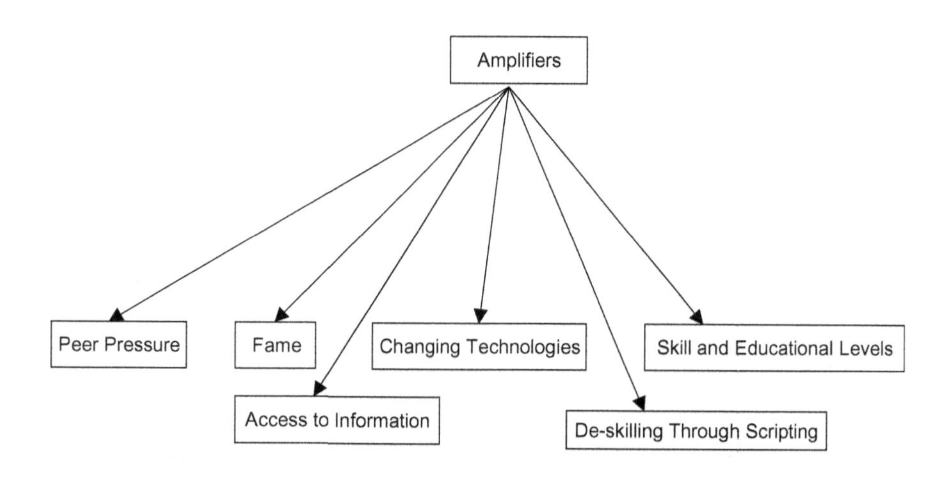

Threat Agents' Capabilities—Nation-States

Factors	Afghanistan	Argentina	Australia
Adult population	14.7 Million[2]	23.5 Million[2] 23.3 Million[8] 23.5 Million[9]	12.9 Million[2] 12.9 Million[9]
Gross domestic product (per capita)	800 US$[2]	7500 US$[8] 7700 US$[9]	22200 US$[2] 23200 US$[8]
Level of literacy	34.4%[2]	96.2%[2] 96.2%[8]	100%[2]
Cultural factors	Sunni muslim – 84%[2]		
Other factors	Major drug production area Support of islamic militants worldwide	Dispute with the United Kingdom over Malvinas money laundering drug transshipment	
Government	Currrent coalition government after period under the Taliban, a self-proclaimed fundamentalist government (dictatorship?)	Democracy	Democracy
Power consumption per capita	15.7 kWh[2]	2081 kWh[2]	9904 kWh[2]
Telecomms infrastructure	2900 Fixed[2] 0 Cellular[2]	7.5 Million Fixed[2] 3 Million Cellular[2]	9.58 Million fixed [2] 6.4 Million cellular [2]
Internet access	(1 ISP)[2]	9,00,000[2]	7.77 Million[2]
Technological developmen level	Low (Industry 15%, services 15% of labour force)		High (industry 22%, services 73% of labour force)
Technical expertise	Very limited	Adequate	High level
Allied nations capability	Limited	High	Very high
Known indigenous IW capability	None	None	Some

Threat Agent Capability—Nation-State (Values for Afghanistan)

Factor	Weighting value				
	1	2	3	4	5
Adult population (P)	< 1,000,000	1,000,001 – 10,000,000	10,000,001 – 50,000,000	50,000,001 – 100,000,000	>100,000,001
Literacy level (L)	< 50%	51–65%	66–80%	80–90%	> 90%
Internet access (I)	Very low	low	Medium	High	Very high
History of relevant activity (H)	None	Intermittent	Occasional	Regular	Regular and widespread
Technical expertise (T)	None	Very limited	Limited	Adequate	High level
Gross domestic product per capita (G)	< $ 1000	$1001 – $5000	$5001– $10,000	$10,001 – $20,000	> $20,000
Allied nation capability (N)	None	Limited	Medium	High	Very high
Indigenous IW capability (AA)	None	Limited	Medium	High	Very high
Other factors (AB)					Religious fundamentalism /svupport of international terrorism (*)

(*) This value is assessed to be of high weighting for this type of threat agent. This is not considered to be a value that would have a lesser importance expression.

Factor	Weighting value				
	1	2	3	4	5
Adult population (P)	<1,000,000	1,000,001 – 10,000,000	10,000,001 – 50,000,000	50,000,001 – 100,000,000	>100,000,001
Literacy level (L)	<50%	51–65%	66–80%	80–90%	>90%
Internet access (I)	Very low	Low	Medium	High	Very high
History of relevant activity (H)	None	Intermittent	Occasional	Regular	Regular and widespread
Technical expertise (T)	None	Very limited	Limited	Adequate	High level
Gross domestic product per capita (G)	< $ 1,000	$1001 – $5000	$5001 – $10,000	$10,001 – $20,000	>$20,000
Allied nation capability (N)	None	Limited	Medium	High	Very high
Indigenous IW capability (AA)	None	Limited	Medium	High	Very high
Other factors (AB)					Religious fundamentalism/ support of international terrorism (*)

Threat Amplifiers

Factors	Environment	Threat agent	Target
Peer pressure		X	
Fame		X	X
Access to information		X	
Changing technologies	X		X
Skills and education levels		X	
De-skilling through scripting		X	
Law enforcement activity (LE)	X	X	X
Target vulnerability (TV)			X
Target profile (TP)		X	X
Public perception (PuP)	X	X	X
Peer perception (PP)		X	

Threat Agent Motivation—General

Factors	State sponsored	Terrorist	Criminal	Pressure group	Commercial	Hacker	Disaffected staff
Personal gain:							
Revenge			X		X	X	X
Power						X	X
Curiosity						X	
Financial			X	X	X	X	X
Peer Recognition						X	
Knowledge or information			X	X	X	X	X
Competitive Advantage			X		X		
Crime		X	X				X
Secular influence		X		X		X	X
National political/ Military objectives	X	X		X			
Religious influence	X	X		X		X	
Pressure group action		X		X			

Threat Agent Motivation—Terrorist (Values for Al Qaeda)

Factor	Weighting value				
	0	12	25	37	50
Crime	No criminal influence	Occasional criminal involvement	Regular criminal involvement	Strong criminal connections	Criminal backing or reliance on funds from criminal activity
Secular influence	No secular influence	Slight secular influence	Moderate secular influence	Strong secular influence	Overriding secular influence
National political/ military objectives	No political or military objectives	Limited political objectives	Moderate political objectives	Political or limited military objectives	Military or strong political objectives
Religious influence	No religious influence	Slight religious influence	Moderate religious influence	Strong religious influence	Overriding religious influence
Pressure group action	No pressure group connection	Slight pressure group connections	Moderate pressure group connections	Strong pressure group relationship	Overriding pressure group relationship

Threat Amplifiers—Terrorists

Factor	Weighting value				
	0	4	8	11	15
Access to information	Not interested in gaining information	Information is only of peripheral interest	Information is a secondary benefit	Information is seen to be of significant interest	Primary reason for attack is to gain information
Changing technologies	Not providing any new opportunity	Potential new opportunities	Limited new opportunities	Providing significant new opportunities	Providing major new opportunities
Law enforcement activity (LE)	Strong and active law enforcement activity in target or base country	Laws in place and enforced with reasonable success in target or base country	Laws in place and enforced with limited success in target or base country	Laws in place but not actively enforced in target or base country	No effective law enforcement activity in target or base country
Target vulnerability (TV)	Target not perceived to be accessible	Target accessible with considerable effort and/or resources	Target accessible with reasonable effort and/or resources	Target accessible with limited effort and/or resources	Target is extremely vulnerable
Target profile (TP)	Does not match group requirement	Slightly matches group requirement	Partially matches group requirement	Mostly matches group requirement	Fully matches group requirement
Fame	Group actively does not wish to be attributed with activity	Group prefers not to be attributed with activity	Group is indifferent to being attributed with the activity	Group is content to be attributed with the activity	Group actively seeks to be attributed with the activity
Public perception (PuP)	Strong negative effect	Negative effect	Neutral effect	Beneficial effect	Strong beneficial effect

Threat Inhibitors

Factors	Environment	Threat agent	Target
Fear of capture		X	
Fear of failure		X	
Level of technical difficulty	X		X
Cost of participating		X	
Sensitivity to public opinion		X	
Law enforcement activity	X	X	
Security of target			X
Public perception		X	X
Security of system		X	X

Threat Inhibitors—Terrorist

Factor	Weighting Value				
	0	4	8	12	16
Fear of capture	Relishes capture – martyrdom?	Seeks capture	Indifferent to capture	Avoids capture	Capture not an option
Fear of failure	Failure would not have any effect	Failure would have a minimal effect	Failure would have an impact but can be accepted	Failure would have a negative impact on the group	Failure would severely damage image or capability
Level of technical difficulty	Easy to obtain success	Success achievable with limited effort and/or resources	Success achievable with reasonable effort and/or resources	Success achievable with significant effort and/or resources	Extremely difficult to achieve success
Sensitivity to public perception	Strong beneficial effect	Beneficial effect	Neutral effect	Negative effect	Strong negative effect
Law enforcement activity	No effective law enforcement activity in target or base country	Laws in pace but not actively enforced in target or base country	Laws in place and enforced with limited success in target or base country	Laws in place and enforced with reasonable success in target or base country	Strong and active law enforcement activity in target or base country
Security of target	Target unprotected	Poor security	Reasonable security	Good security	Target extremely secure

Threat Amplifiers

Factors	Environment	Threat agent	Target
Peer pressure		X	
Fame		X	X
Access to information		X	
Changing technologies	X		X
Skills and education levels		X	
De-skilling through scripting		X	
Law enforcement activity (LE)	X	X	X
Target vulnerability (TV)			X
Target profile (TP)		X	X
Public perception (PuP)	X	X	X
Peer perception (PP)		X	

Threat Amplifiers—Terrorists

Factor	Weighting value				
	0	4	8	11	15
Access to information	Not interested in gaining information	Information is only of peripheral interest	Information is a secondary benefit	Information is seen to be of significant interest	Primary reason for attack is to gain information
Changing technologies	Not providing any new opportunity	Potential new opportunities	Limited new opportunities	Providing significant new opportunities	Providing major new opportunities
Law enforcement activity (LE)	Strong and active law enforcement activity in target or base country	Laws in place and enforced with reasonable success in target or base country	Laws in place and enforced with limited success in target or base country	Laws in place but not actively enforced in target or base country	No effective law enforcement activity in target or base country
Target vulnerability (TV)	Target not perceived to be accessible	Target accessible with considerable effort and/or resources	Target accessible with reasonable effort and/or resources	Target accessible with limited effort and/or resources	Target is extremely vulnerable
Target profile (TP)	Does not match group requirement	Slightly matches group requirement	Partially matches group requirement	Mostly matches group requirement	Fully matches group requirement
Fame	Group actively does not wish to be attributed with activity	Group prefers not to be attributed with activity	Group is indifferent to being attributed with the activity	Group is content to be attributed with the activity	Group actively seeks to be attributed with the activity
Public perception (PuP)	Strong negative effect	Negative effect	Neutral effect	Beneficial effect	Strong beneficial effect

Catalysts

Factors	
Change of personal circumstances	X
War or political conflict	X
Significant events	X
Significant anniversaries	X

Terrorists Threat Agents—Catalysts

Factor	Weighting value				
	0	8	16	24	32
War or political conflict	No current political or military hostilities	Low level of hostile political activity	Medium level of hostile political activity	High level of political activity and potential for military action	Currently directly involved in a war that is taking place
Significant events	No significant events	Awareness of event of limited peripheral significance	Awareness of event of minor significance	Awareness of event of significance	Awareness of event of major significance
Significant anniversaries	No anniversary of significance to either the threat agent or the target	Anniversary of limited relevance to the threat agent or the target	Anniversary of relevance to the threat agent or the target	Anniversary of some significance to the threat agent or the target	Anniversary of high significance to the threat agent or the target

Where does all this lead?

	Capability	Motivation	Access	Inhibitors	Amplifiers	Catalysts
Nation State (Afghanistan)	24	62	45	40	50	92
Terrorism (Al Qa'eda)	47	62	42	34	50	67
Crime (Cali Cartel)	77	45	54	59	47	25
Pressure Group (ALF)	82	39	38	57	64	78
Hackers (Silver Lords)	85	37	53	42	83	83

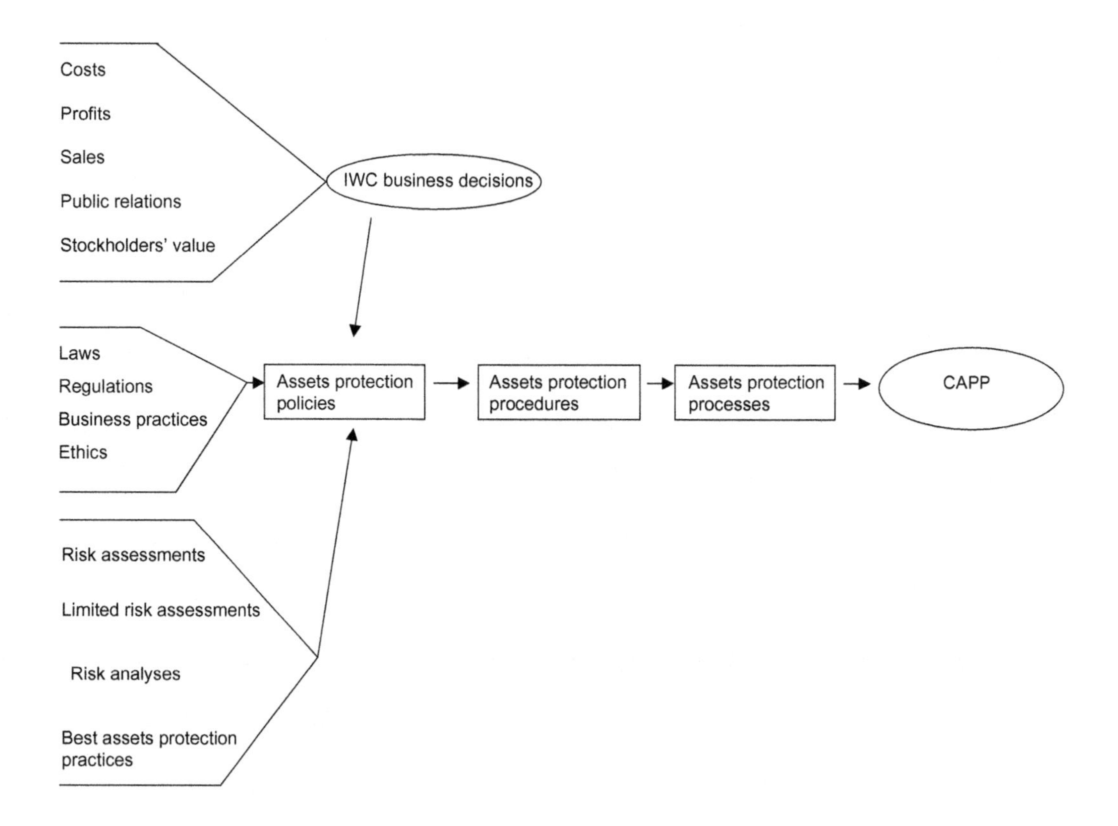
Costs
Profits
Sales
Public relations
Stockholders' value
IWC business decisions
Laws
Regulations
Business practices
Ethics
Assets protection policies
Assets protection procedures
Assets protection processes
CAPP
Risk assessments
Limited risk assessments
Risk analyses
Best assets protection practices

Requirement-driver

- Federal laws and regulations (SEC, etc.)
- Obligations to stockholders (owners)

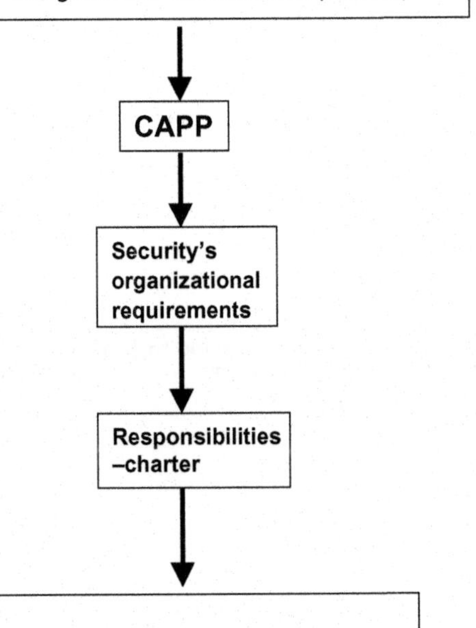

Security organizational functions

- Access controls
- Disaster recovery/emergency planning
- Risk management
- Physical security
- etc.

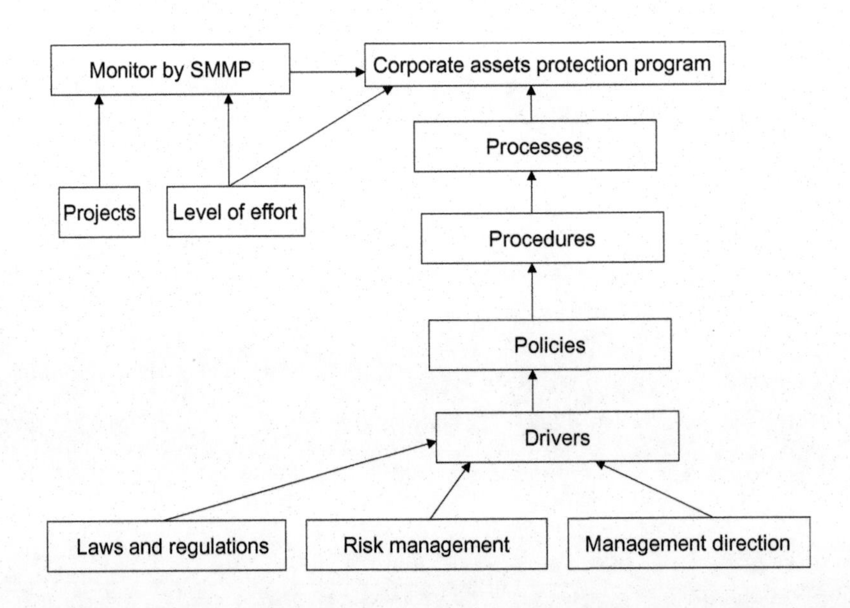

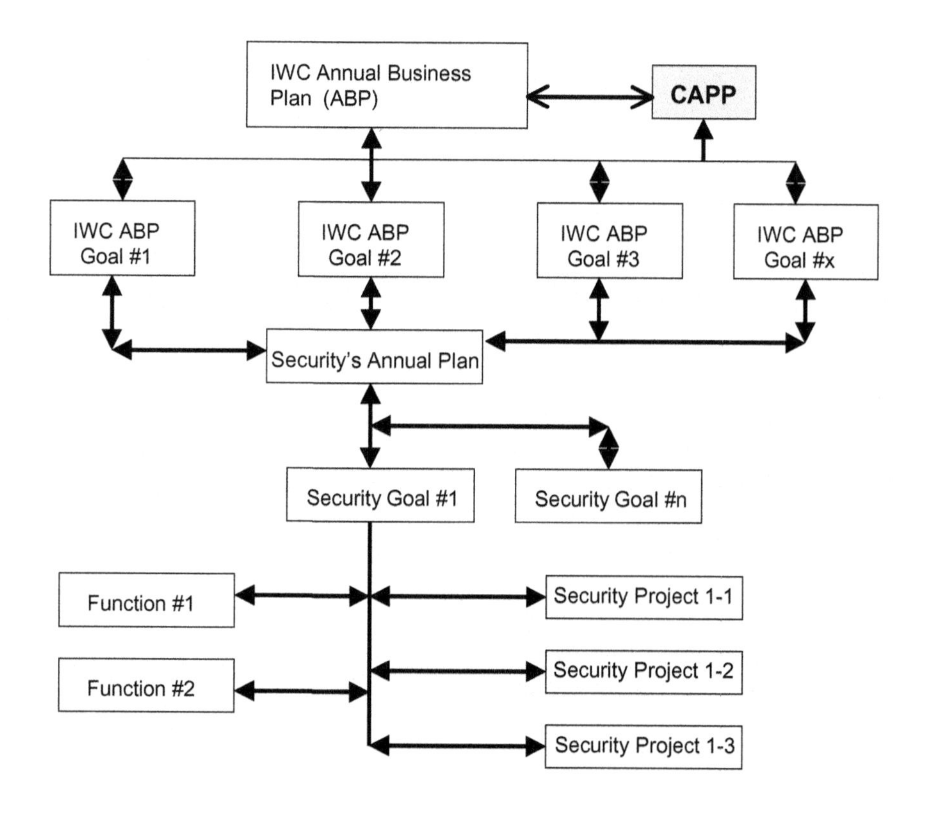
IWC Annual Business Plan (ABP)
CAPP
IWC ABP Goal #1
IWC ABP Goal #2
IWC ABP Goal #3
IWC ABP Goal #x
Security's Annual Plan
Security Goal #1
Security Goal #n
Function #1
Function #2
Security Project 1-1
Security Project 1-2
Security Project 1-3

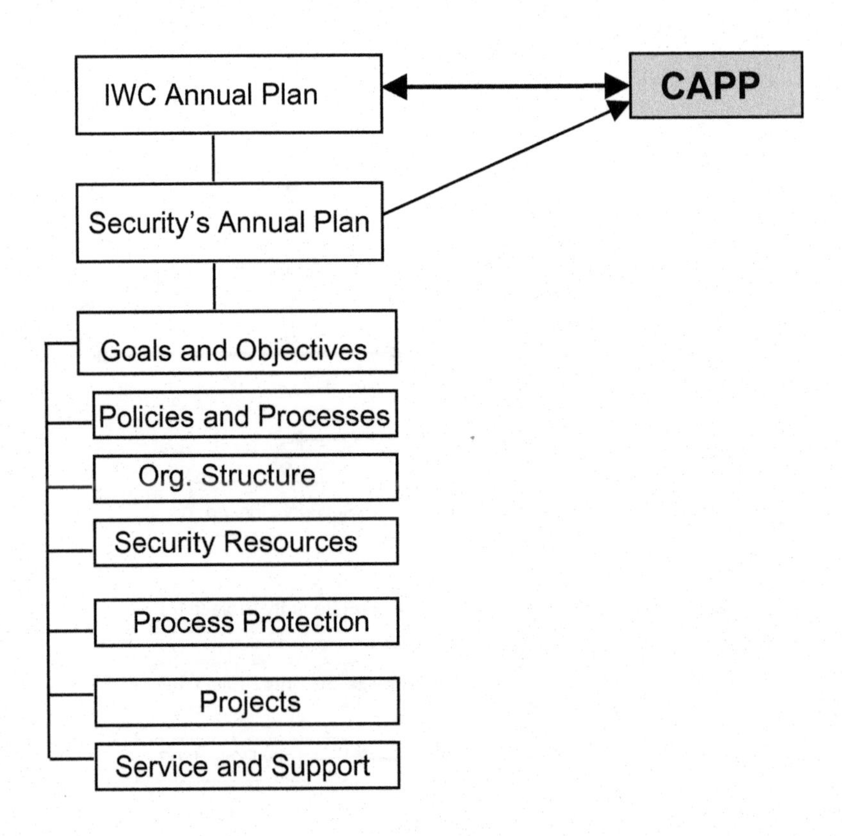
IWC Annual Plan
CAPP
Security's Annual Plan
Goals and Objectives
Policies and Processes
Org. Structure
Security Resources
Process Protection
Projects
Service and Support

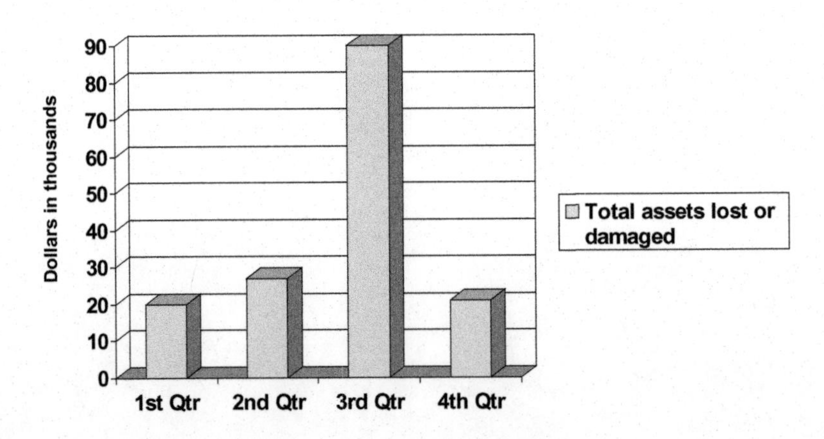
IWC Total Assets Lost or Damaged in 2005
Dollars in thousands
90
80
70
60
50
40
30
20
10
0
1st Qtr
2nd Qtr
3rd Qtr
4th Qtr
Total assets lost or damaged

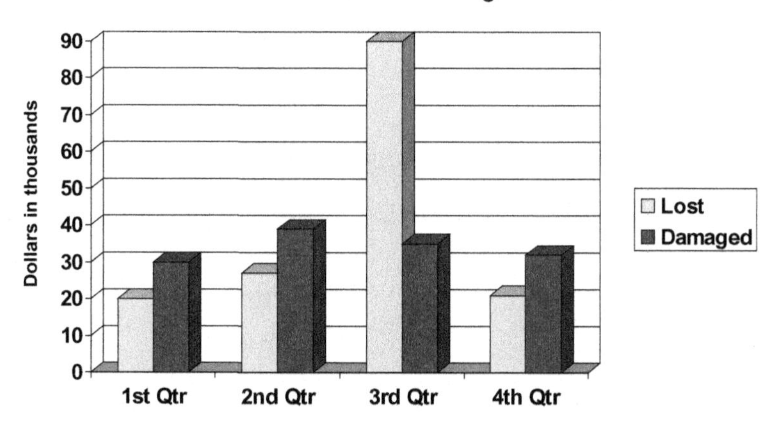
IWC Assets Lost vs Damaged—2005
Dollars in thousands
90
80
70
60
50
40
30
20
10
0
1st Qtr
2nd Qtr
3rd Qtr
4th Qtr
Lost
Damaged

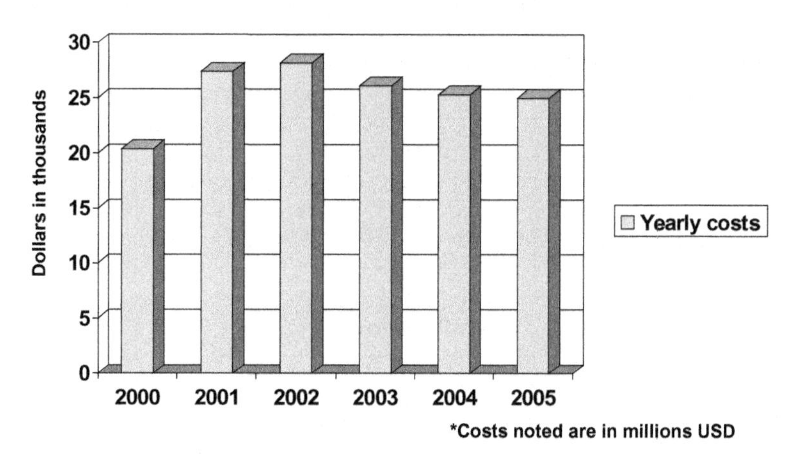
IWC CAPP's Total Costs 2000–2005
Dollars in thousands
30
25
20
15
10
5
0
2000
2001
2002
2003
2004
2005
Yearly costs
*Costs noted are in millions USD

IWC CAPP's Total Costs 2000–2005

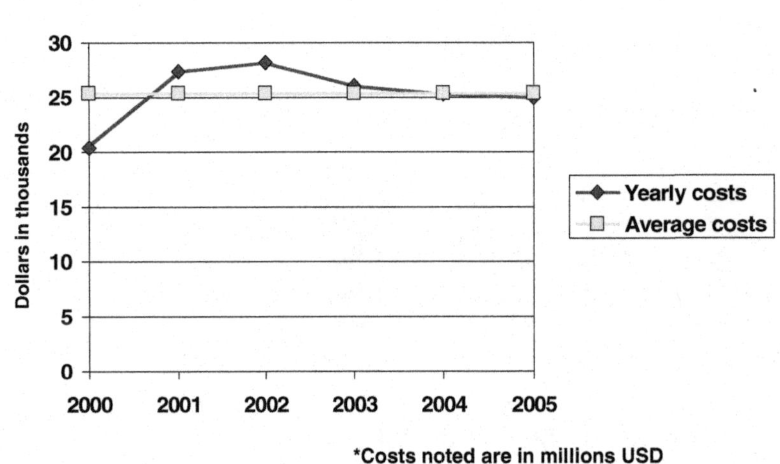

- 2000: Business without a CAPP and heavy losses
- 2001: Ramping up to a cost-effective CAPP
- 2002: Ramping up to a cost-effective CAPP
- 2003: CAPP costs downward trend due to maturity and some efficiency gains
- 2004: CAPP costs continue downward trend due to efficiency gains
- 2005: CAPP costs continue downward trend due to efficiency gains and now less than 6-year average costs

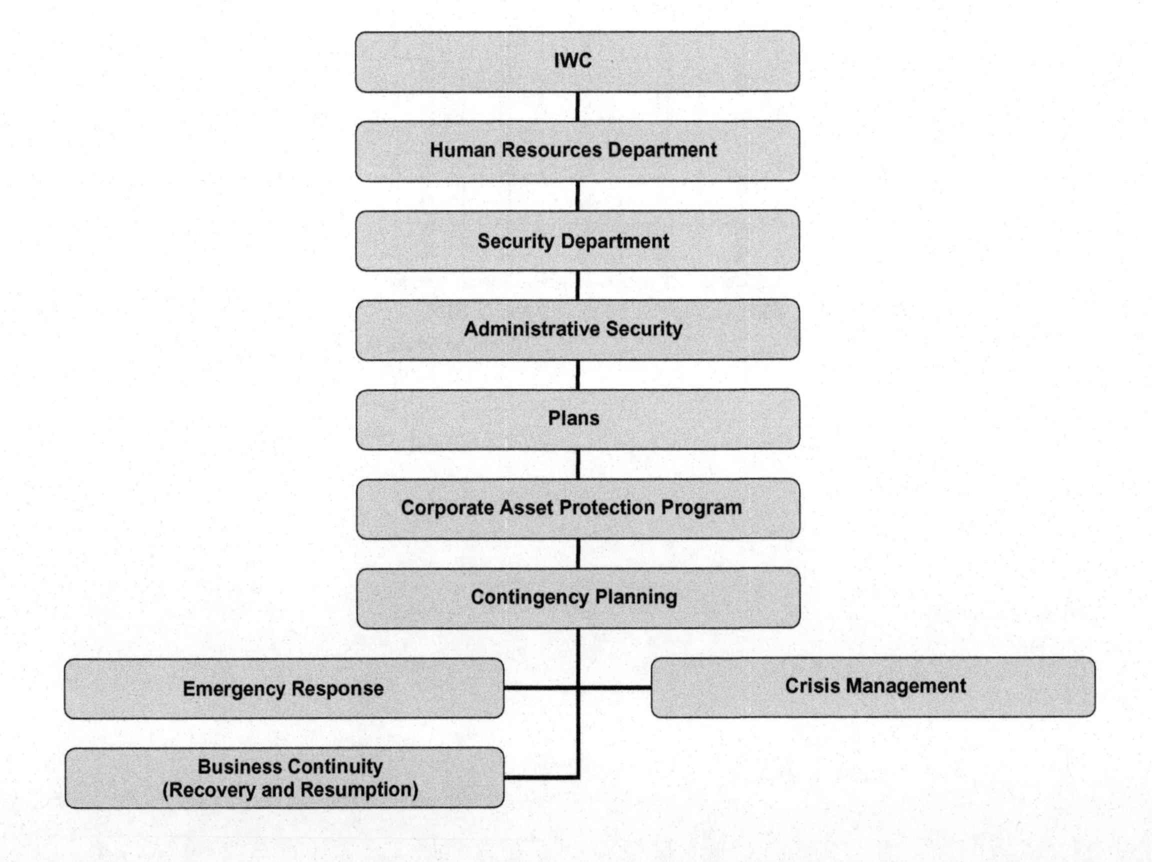

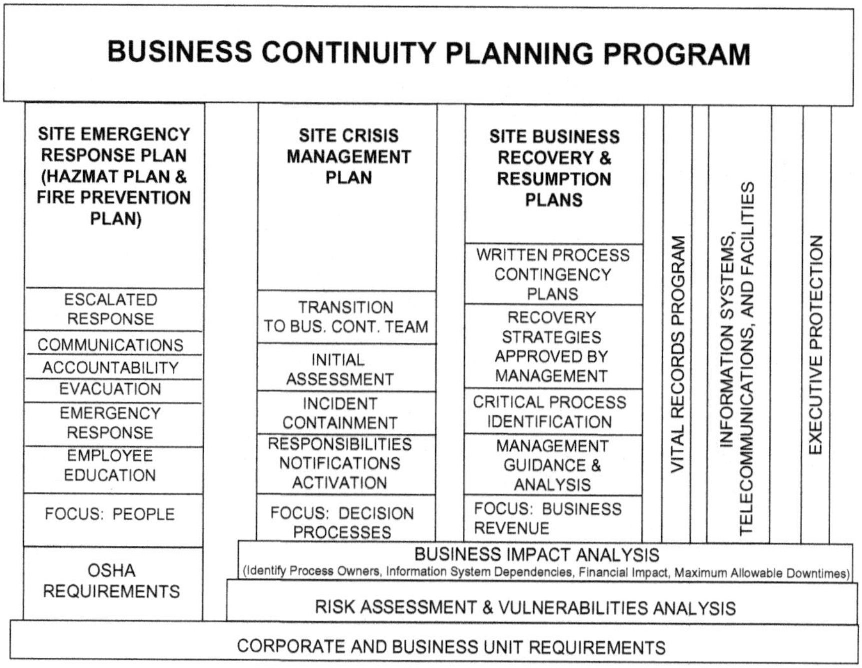

BUSINESS CONTINUITY PLANNING PROGRAM

SITE EMERGENCY RESPONSE PLAN (HAZMAT PLAN & FIRE PREVENTION PLAN)
ESCALATED RESPONSE
COMMUNICATIONS
ACCOUNTABILITY
EVACUATION
EMERGENCY RESPONSE
EMPLOYEE EDUCATION
FOCUS: PEOPLE
OSHA REQUIREMENTS

SITE CRISIS MANAGEMENT PLAN
TRANSITION TO BUS. CONT. TEAM
INITIAL ASSESSMENT
INCIDENT CONTAINMENT
RESPONSIBILITIES NOTIFICATIONS ACTIVATION
FOCUS: DECISION PROCESSES

SITE BUSINESS RECOVERY & RESUMPTION PLANS
WRITTEN PROCESS CONTINGENCY PLANS
RECOVERY STRATEGIES APPROVED BY MANAGEMENT
CRITICAL PROCESS IDENTIFICATION
MANAGEMENT GUIDANCE & ANALYSIS
FOCUS: BUSINESS REVENUE

VITAL RECORDS PROGRAM
INFORMATION SYSTEMS, TELECOMMUNICATIONS, AND FACILITIES
EXECUTIVE PROTECTION

BUSINESS IMPACT ANALYSIS
(Identify Process Owners, Information System Dependencies, Financial Impact, Maximum Allowable Downtimes)
RISK ASSESSMENT & VULNERABILITIES ANALYSIS
CORPORATE AND BUSINESS UNIT REQUIREMENTS

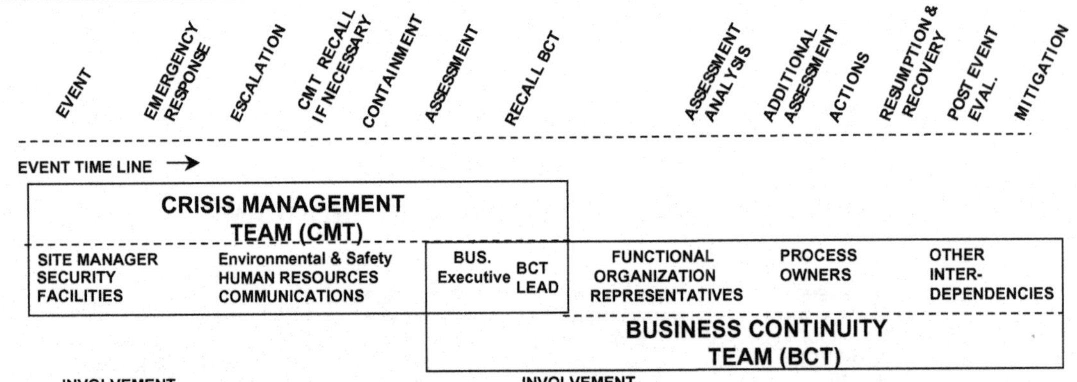

INVOLVEMENT
- EMERGENCY RESPONSE/CRISIS MANAGEMENT PLANNING PROCESSES
- DEVELOPS SITE RISK AND VULNERABILITIES ANALYSIS
- IDENTIFICATION OF POTENTIAL MITIGATION DECISIONS/ACTIVITIES

RESULTS
- OSHA COMPLIANCE
- EMERGENCY PREPAREDNESS AND EMPLOYEE PROTECTION
- EMERGENCY RESPONSE/INCIDENT MANAGEMENT
- EMERGENCY AND CRISIS MANAGEMENT PLANS

INVOLVEMENT
- REVIEW OF CMT GENERATED RISK AND VULNERABILITIES ANALYSIS
- BUSINESS IMPACT ANALYSIS
- IDENTIFICATION OF CRITICAL PROCESSES/SYSTEMS RESULTING IN APPROVED CRITICAL PROCESS LISTS (CPL)
- MITIGATION DECISIONS/ACTIVITIES

RESULTS
- TARGETED EXPENDITURES SUPPORT CRITICAL PROCESSES/REVENUE OPERATIONS
- MANAGEMENT OF BUSINESS RESUMPTION PLANNING
- PROCESS CONTINGENCY PLANS FOR CRITICAL PROCESSES
- BUSINESS RESUMPTION PLANS FOR CRITICAL PROCESSES, FACILITY AND/OR COMPANY

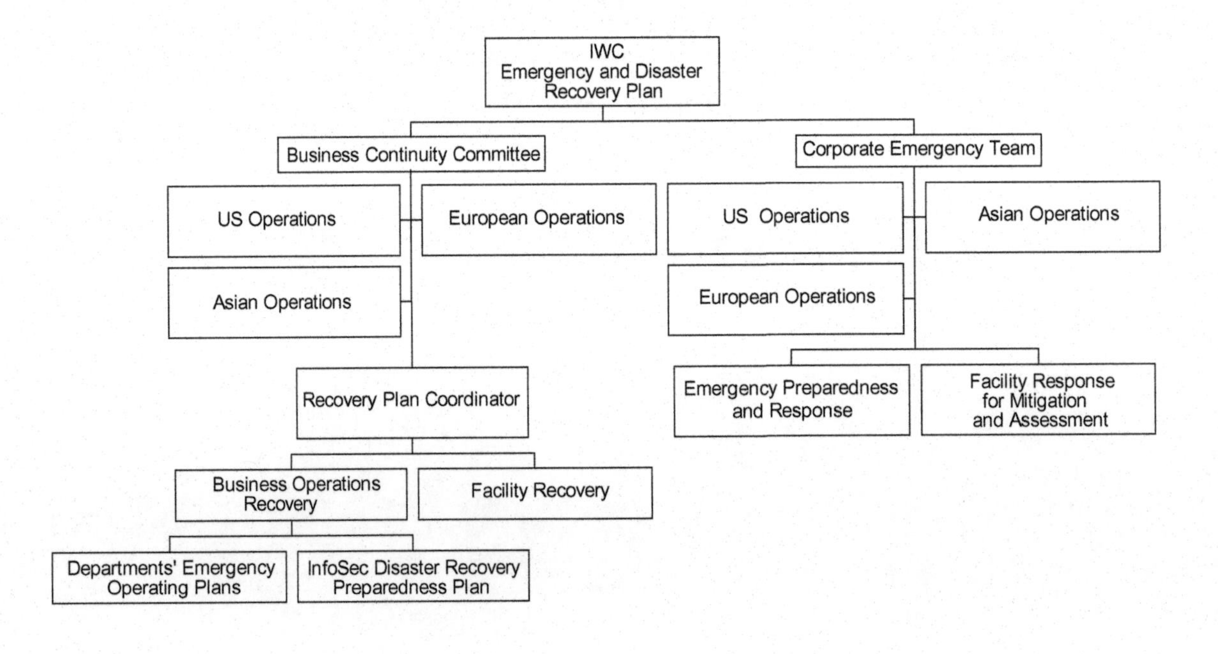

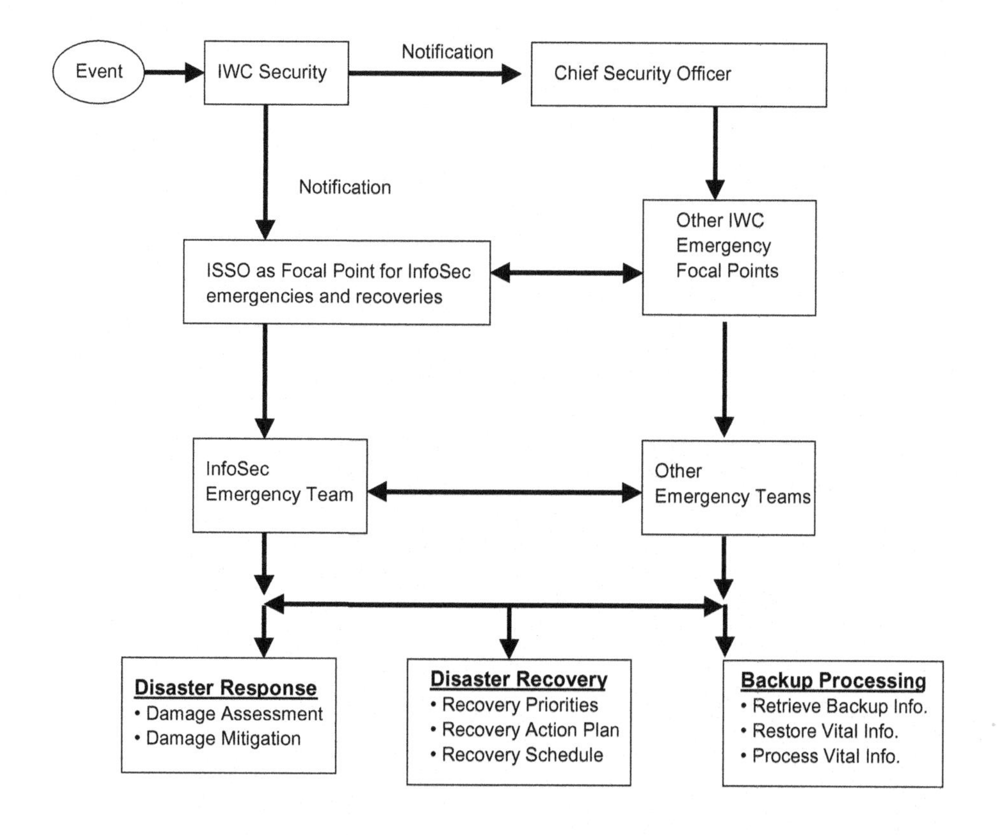

THREAT TO ASSETS IDENTIFICATION MATRIX

	Assets							
Threats	Data	Software	Hardware	Physical Facilities	Media & Supplies	Telecom	Employees	Other
Windstorm								
Snowstorm								
Earthquake								
Volcano								
Landslide								
Minor Fire								
Major Fire								
Catastrophic Fire								
Liquid leakage								
Environmental Problem								
Environmental Interruptions								
Telecom Interruptions								
Power Interruptions								
Power Fluctuations								
Human Errors								
System Errors								
Software Errors								
Data Errors								
Unauthorized Use of Assets								
Fraud								
Sabotage								
Unauthorized Disclosure								
Theft								
Denial of Service								
Others								

Time-Motion Study Collection Sheet

This sheet is to be used to record the individual units of a particular measured task.
It is to be used to track the time it takes to perform contingency planning functions that are part of your responsibilities under the contingency planning function.
A task identifier code will be used that coincides with the numbered task of the contingency planning detailed flowchart.
Cost at $15USD per hour based on average IWC salary and benefits costs.

Task Identifier		Start Time		End Time		Total Time		Cost (USD)

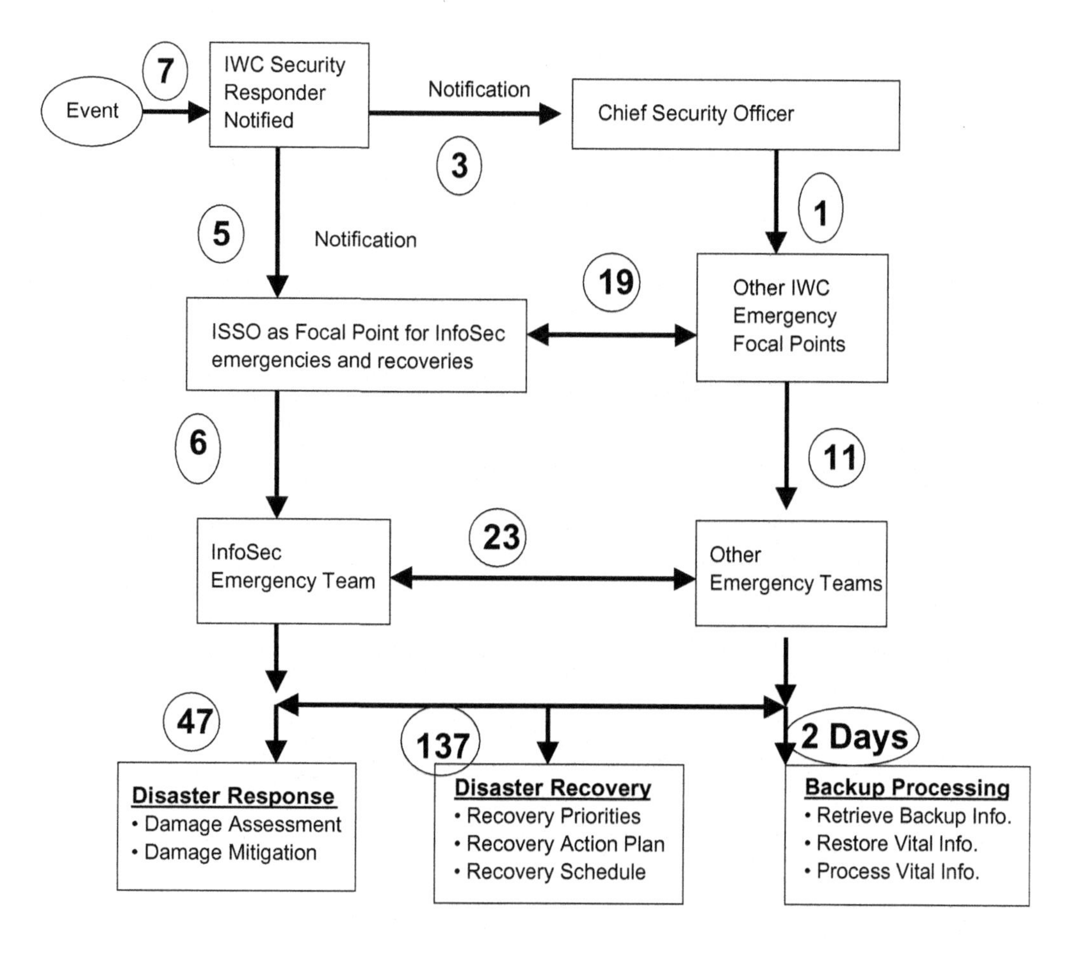

NOTE: All numbers are times in minutes, except where otherwise noted.

Contingency Planning Tests—Costs of Testing

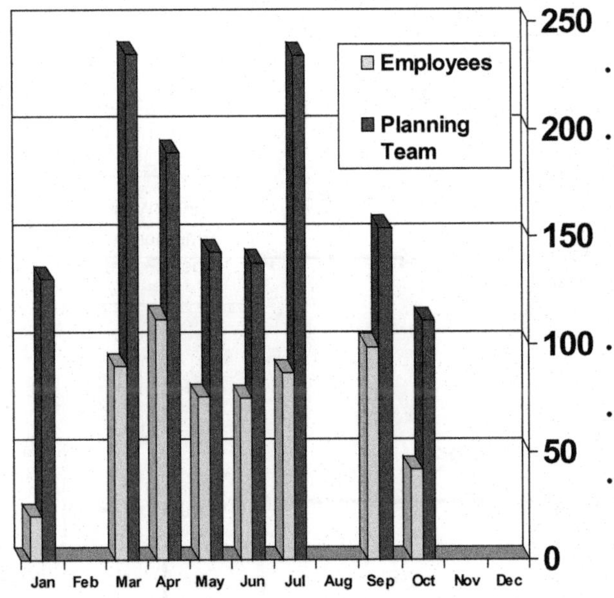

- Eight contingency planning tests were conducted in 2005.
- They tested various aspects of the contingency plans to include use of information backup media, office fires, workplace violence response, earthquake, and terrorists car bomb attack.
- Lost productivity of employees involved were tracked.
- Lost productivity of those affected by the tests were tracked.
- Costs are in hours lost.

Contingency Planning—Tests Results—2005

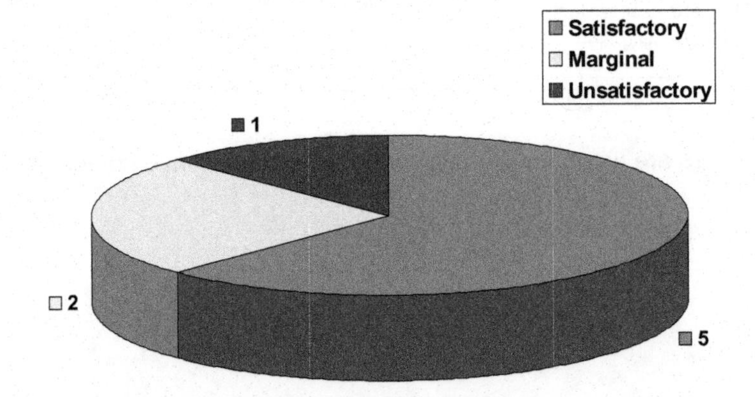

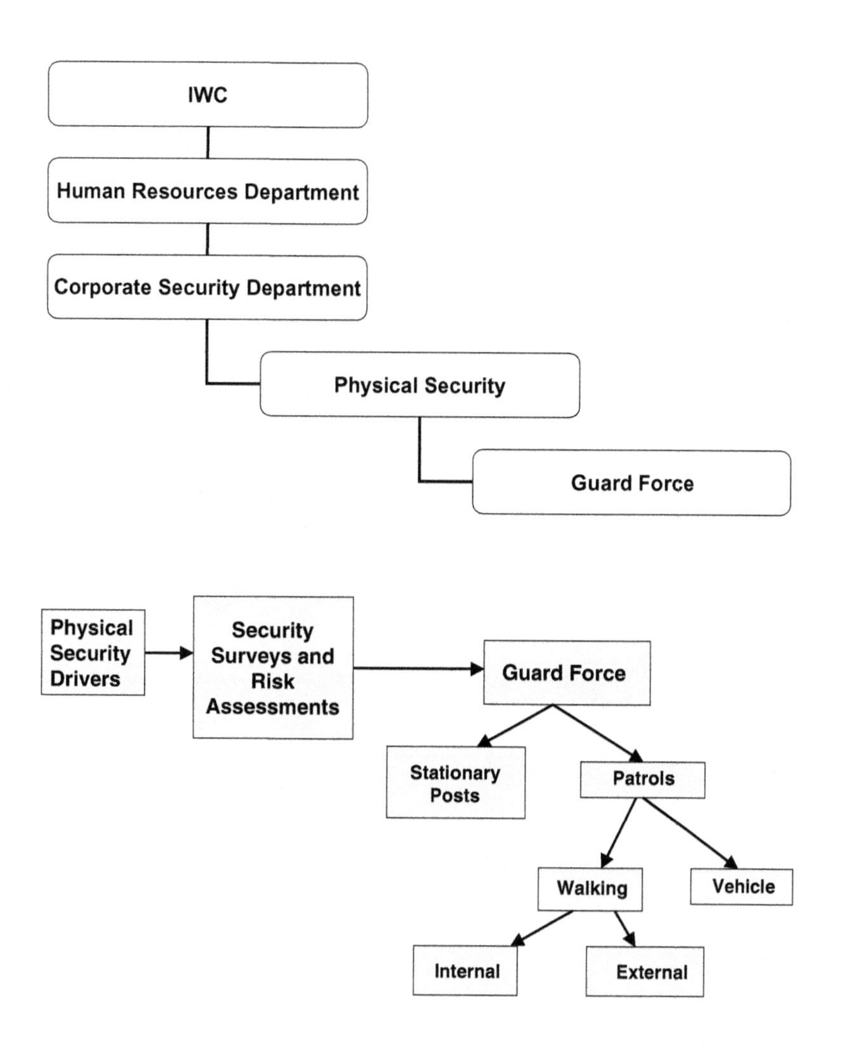
IWC
Human Resources Department
Corporate Security Department
Physical Security
Guard Force
Physical Security Drivers
Security Surveys and Risk Assessments
Guard Force
Stationary Posts
Patrols
Walking
Vehicle
Internal
External

Guard Force: Pre-Posting Meeting Hours Spent—Total Hours for All Guards

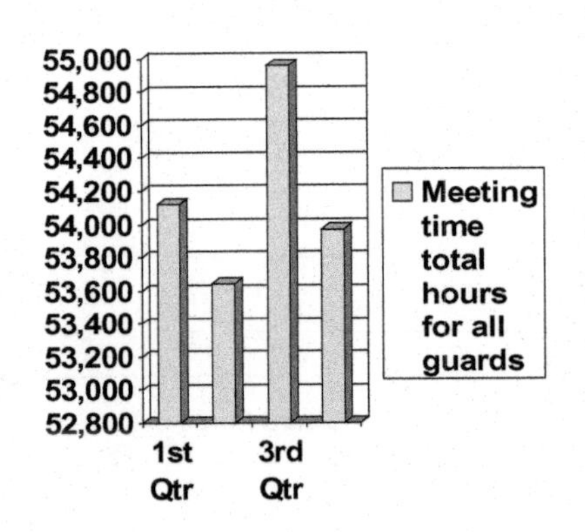

- Objective: Determine if there is a more efficient yet effective way to meet pre-posting meeting goals for the guard force.
- Method: Collect time guards spend in a pre-post meeting.
- Tool: Sign-in and out log: all guards will sign in and out of pre-posting meeting.
- Process: Each guard's time in meeting summed and totaled for each meeting. Then that total is entered into a spreadsheet and summed each quarter and mapped to a bar chart.

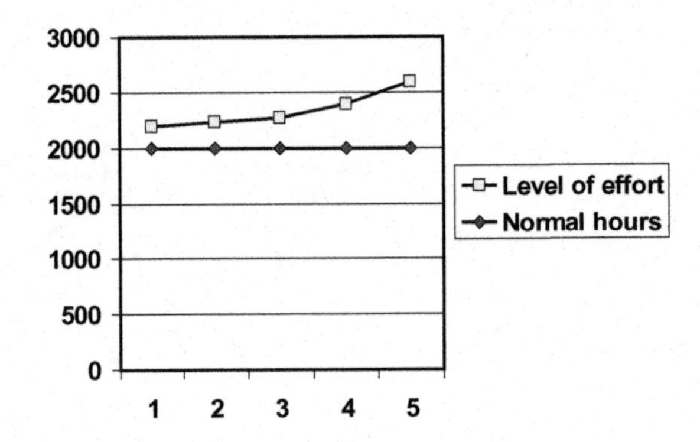

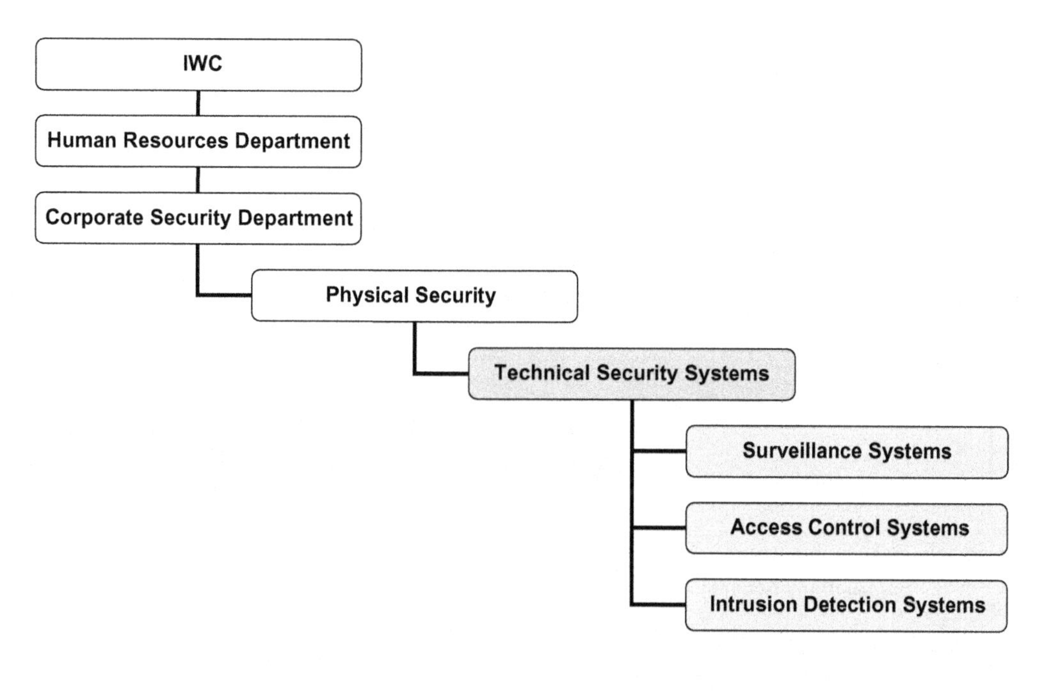
IWC
Human Resources Department
Corporate Security Department
Physical Security
Technical Security Systems
Surveillance Systems
Access Control Systems
Intrusion Detection Systems

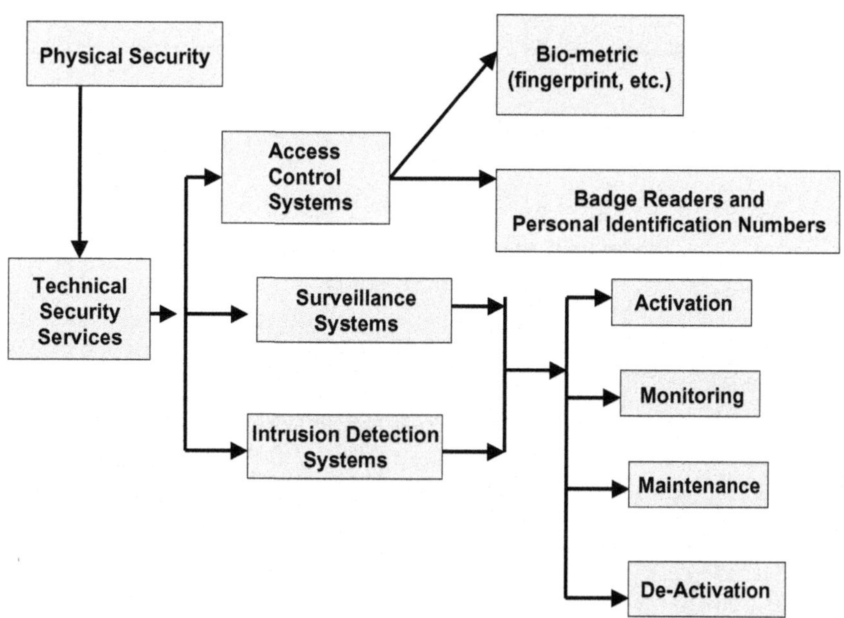
Physical Security
Technical Security Services
Access Control Systems
Bio-metric (fingerprint, etc.)
Badge Readers and Personal Identification Numbers
Surveillance Systems
Intrusion Detection Systems
Activation
Monitoring
Maintenance
De-Activation

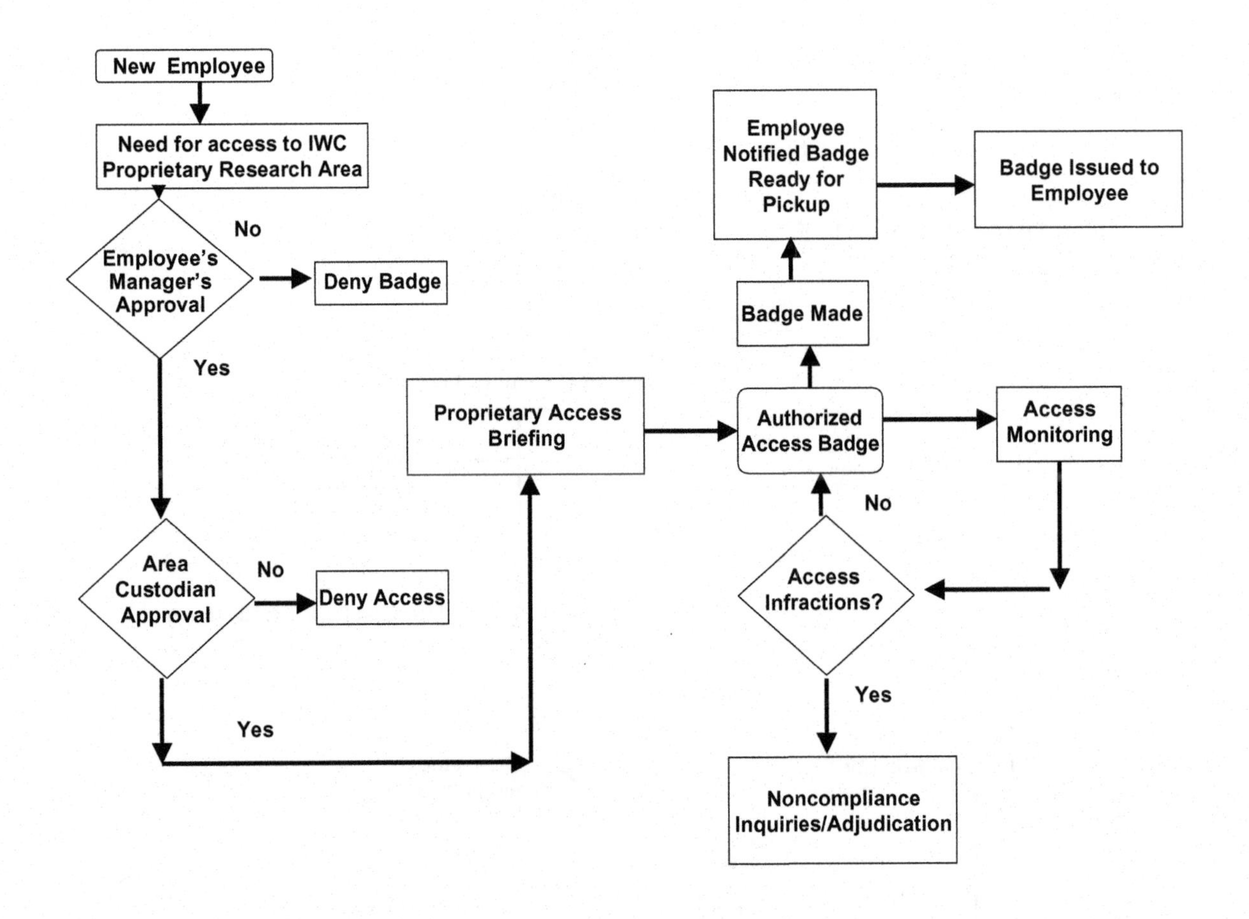

Total IWC Alarm Maintenance Problems Worldwide—2005

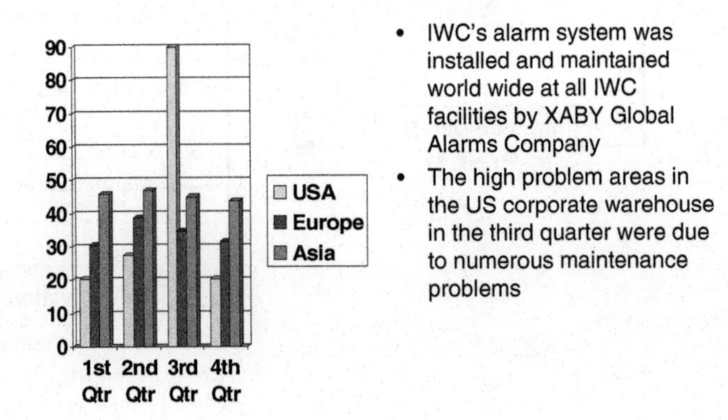

Total IWC Alarm Maintenance Costs Worldwide—2005

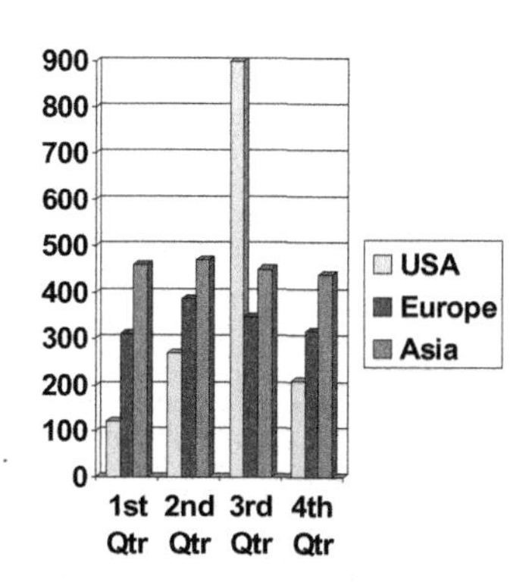

- IWC's alarm system was installed and maintained world-wide at all IWC facilities by XABY Global Alarms Company.
- The costs are the total costs of repair and maintenance at all locations converted into US dollars (thousands).
- The costs include "alarm substitute costs" of using a security guard to replace the alarm system while it is under repair.

Area Access Badge Processing—Average Time in Days

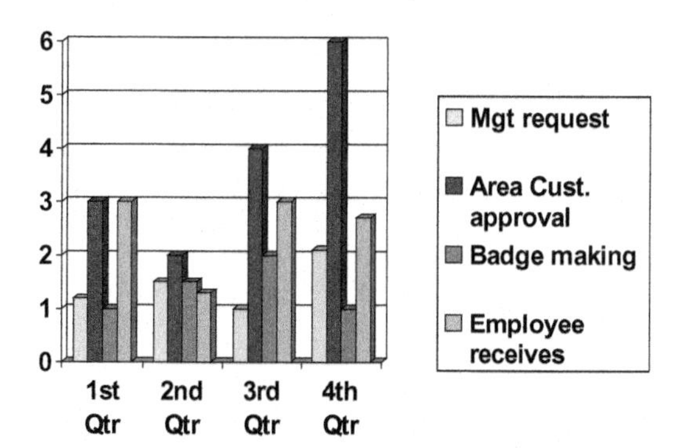

Area Access Badge Processing—
Average Time in Dollars per Badge Processed

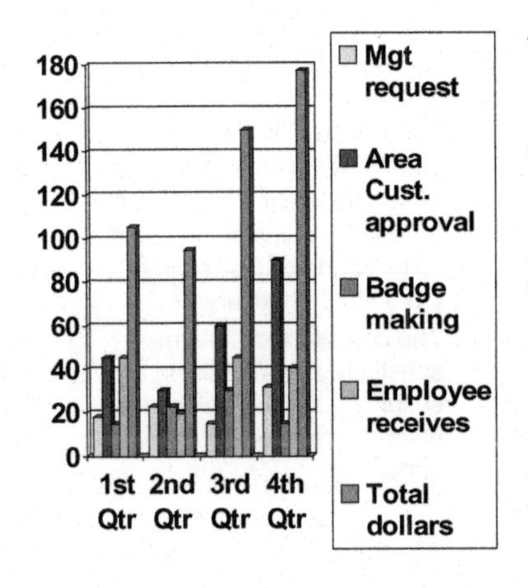

- The area access badge processing time is based on all badge processed on a quarterly basis during 2005.
- The average time was then converted to dollars using $15 per hour, that includes benefits' costs, as provided by the Human Resources staff.
- This is part of a project to identify the time and productivity losses associated with area access badge process and to analyze the data and find ways to improve the process in less time and productivity loss.
- During 2004: 1st Qtr: 27 badges; 2nd Qtr: 39 badges; 3rd Qtr: 15 badges; and 4th Qtr: 12 badges.

Total Area Access Badge Requests for a Year—
Total Costs per Quarter and the Year

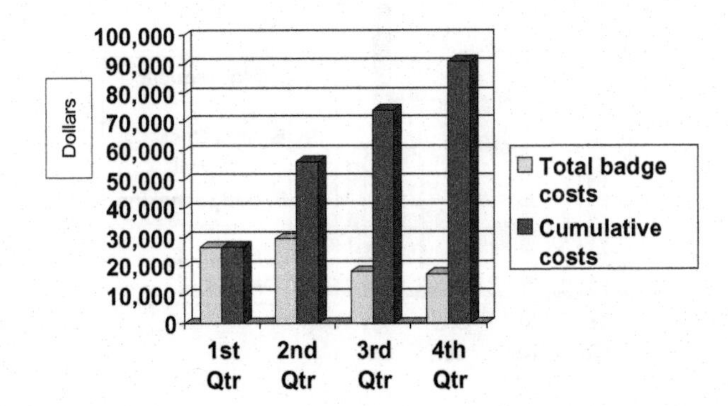

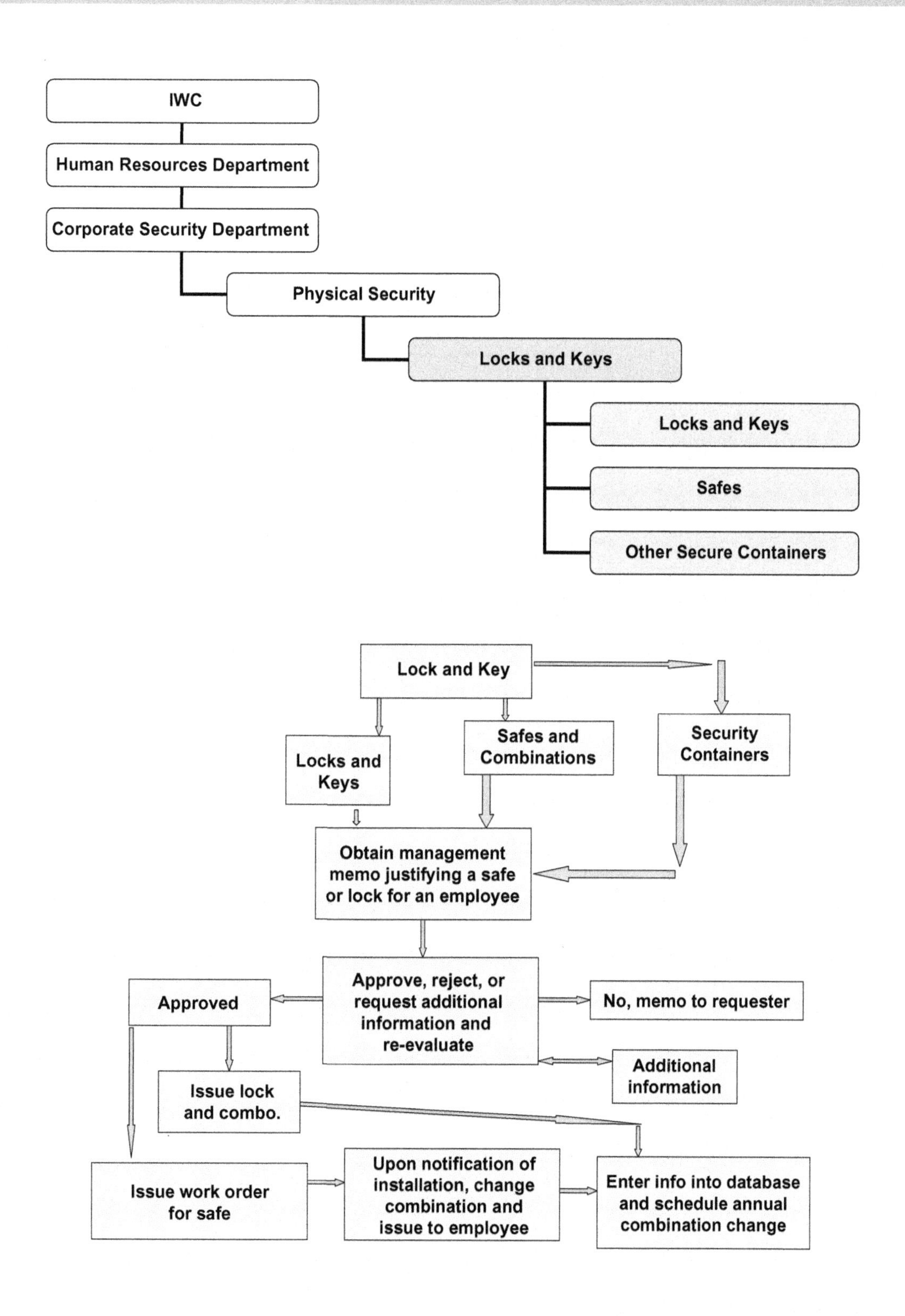
IWC
Human Resources Department
Corporate Security Department
Physical Security
Locks and Keys
Locks and Keys
Safes
Other Secure Containers
Lock and Key
Locks and Keys
Safes and Combinations
Security Containers
Obtain management memo justifying a safe or lock for an employee
Approve, reject, or request additional information and re-evaluate
Approved
No, memo to requester
Additional information
Issue lock and combo.
Issue work order for safe
Upon notification of installation, change combination and issue to employee
Enter info into database and schedule annual combination change

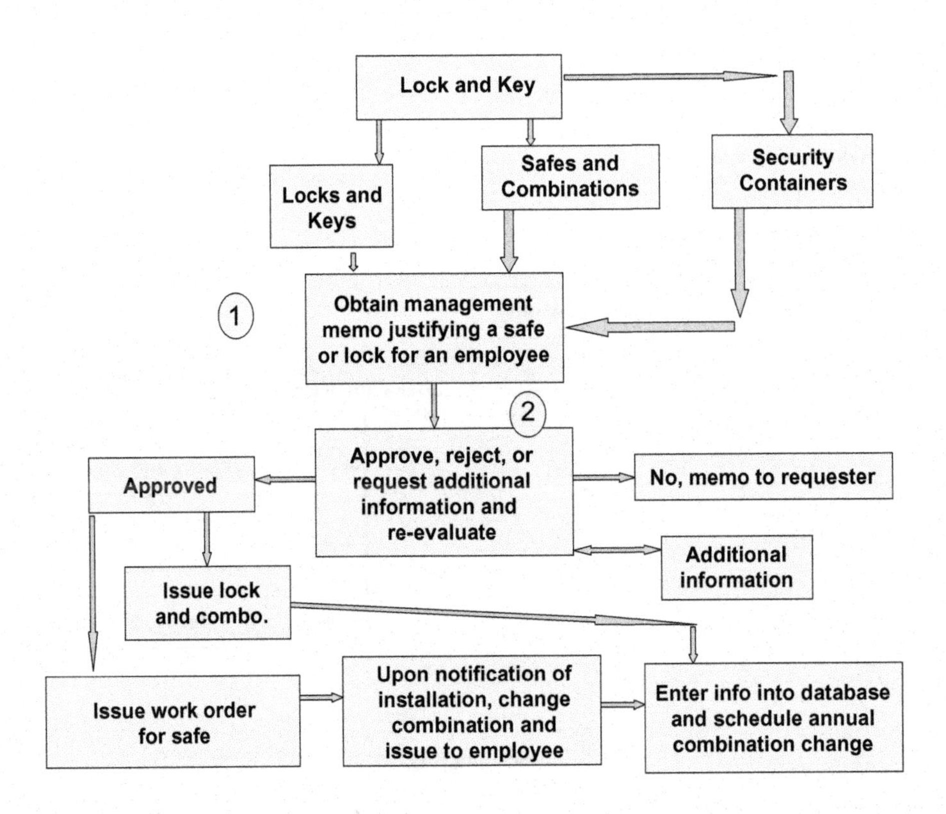

Data Collection Log - Times Between Requests and Deliveries - Lock and Key										
Date	Time In	Time Out	Total Time	Name	Organization	Purpose of Visit	Lock	Key	Secure Cabinet	Safe
TOTALS	Sorted by hour	Sorted by hour	Avg. Time	Sorted by name	Sorted by Org.	Sorted by purpose				
NOTE: Items can be sorted, totaled, sub-totaled as shown by non-white cells										

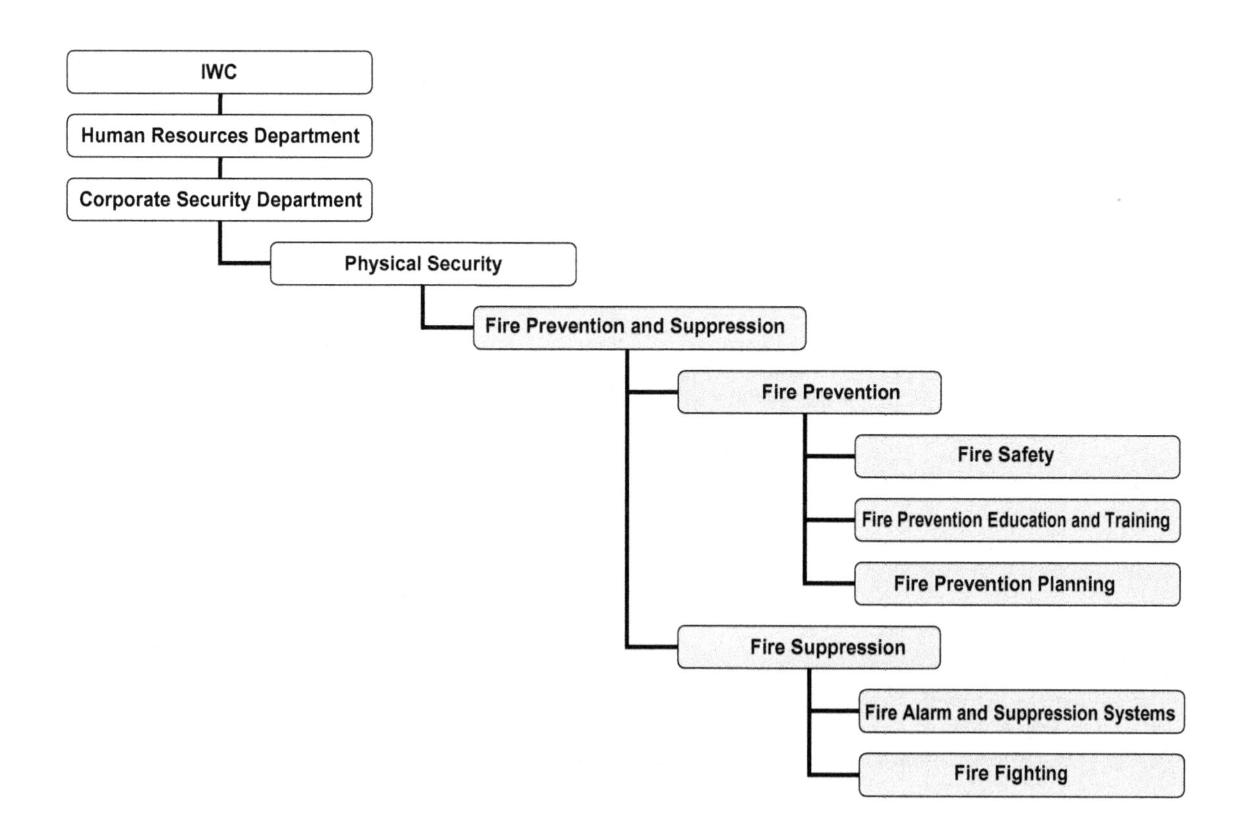
IWC
Human Resources Department
Corporate Security Department
Physical Security
Fire Prevention and Suppression
Fire Prevention
Fire Safety
Fire Prevention Education and Training
Fire Prevention Planning
Fire Suppression
Fire Alarm and Suppression Systems
Fire Fighting

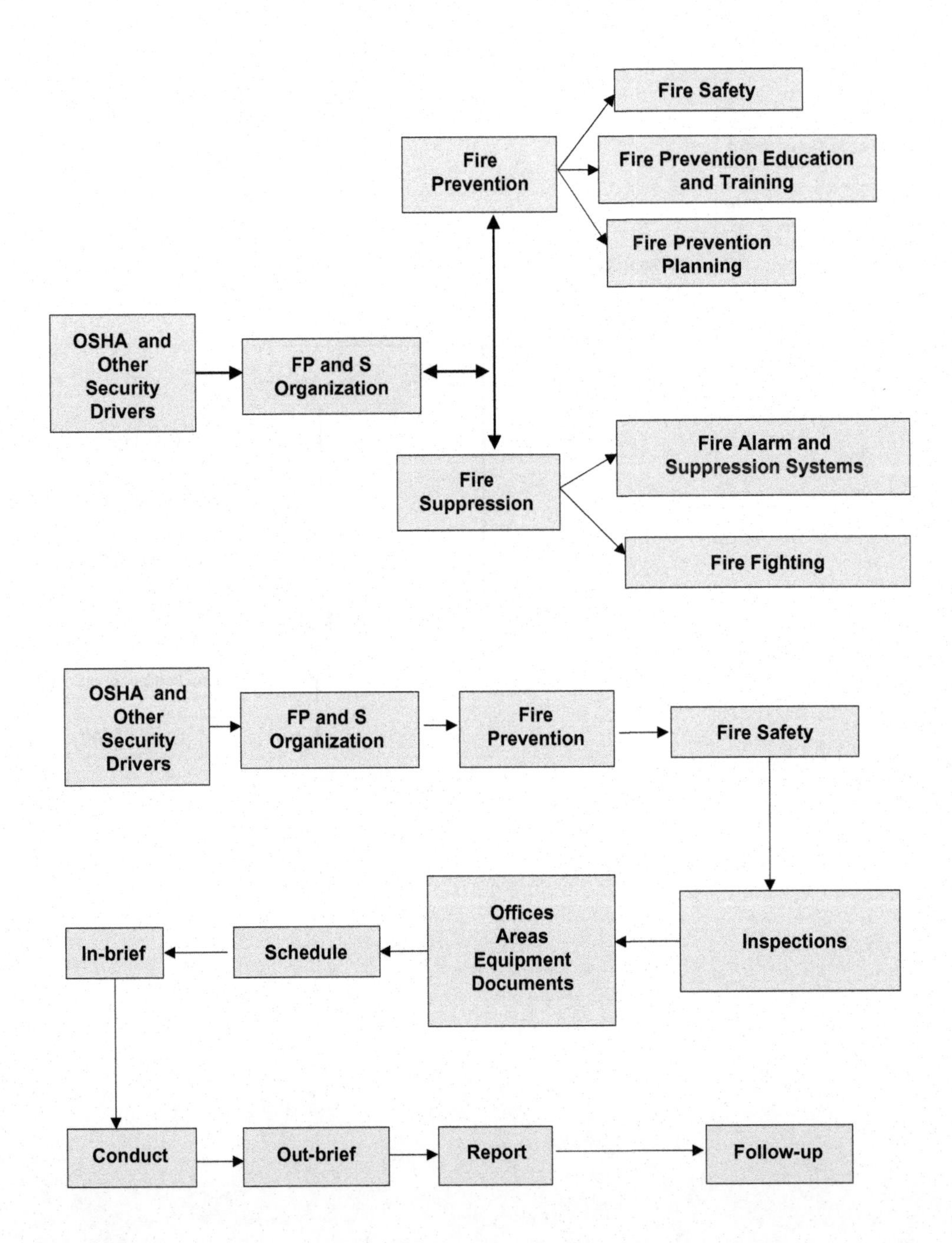
Fire Safety
Fire Prevention
Fire Prevention Education and Training
Fire Prevention Planning
OSHA and Other Security Drivers
FP and S Organization
Fire Alarm and Suppression Systems
Fire Suppression
Fire Fighting
OSHA and Other Security Drivers
FP and S Organization
Fire Prevention
Fire Safety
Offices Areas Equipment Documents
Inspections
In-brief
Schedule
Conduct
Out-brief
Report
Follow-up

IWC—US False Fire Alarms—2005

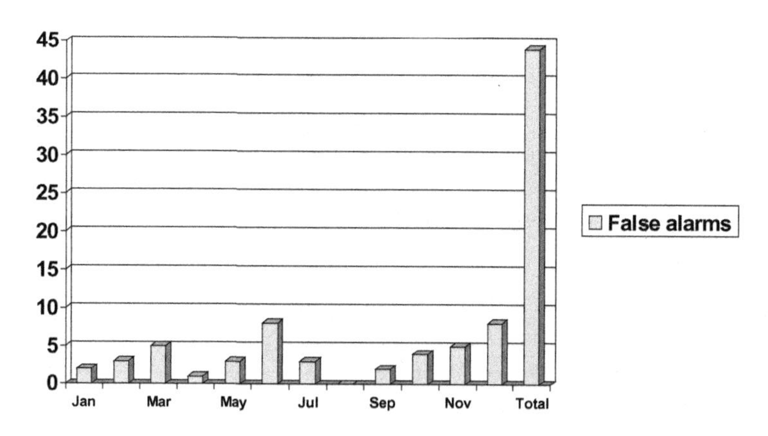

IWC—US Fire Alarm Problem—Maintenance Incidents—2005

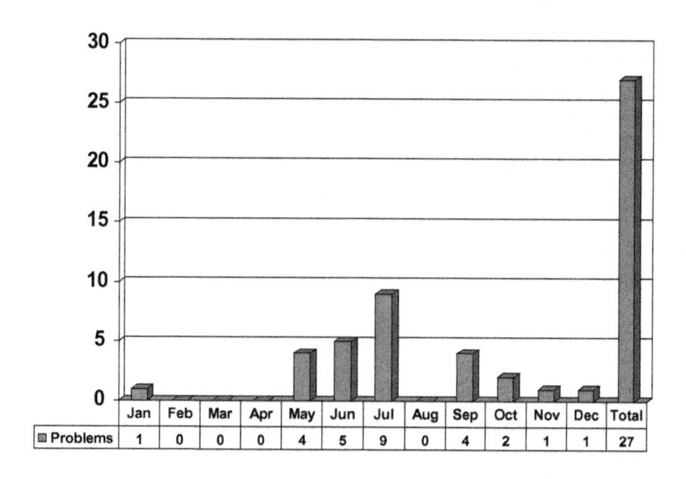

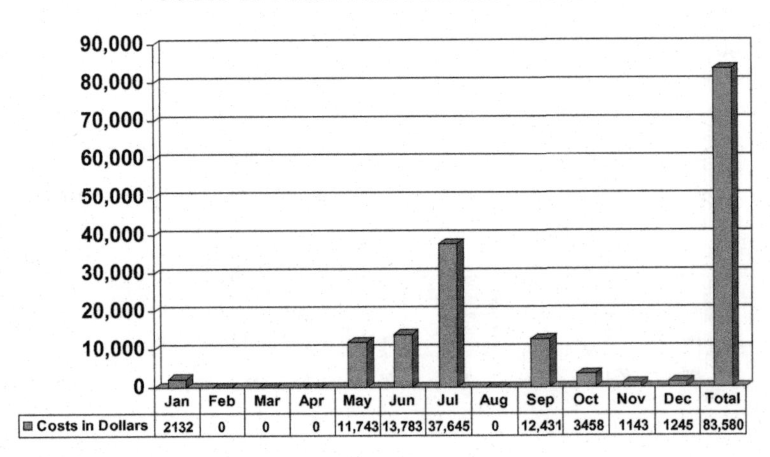
IWC—U.S. Fire Alarm
Costs of False Fire Alarms—2005
90,000
80,000
70,000
60,000
50,000
40,000
30,000
20,000
10,000
0
Jan Feb Mar Apr May Jun Jul Aug Sep Oct Nov Dec Total
Costs in Dollars 2132 0 0 0 11,743 13,783 37,645 0 12,431 3458 1143 1245 83,580

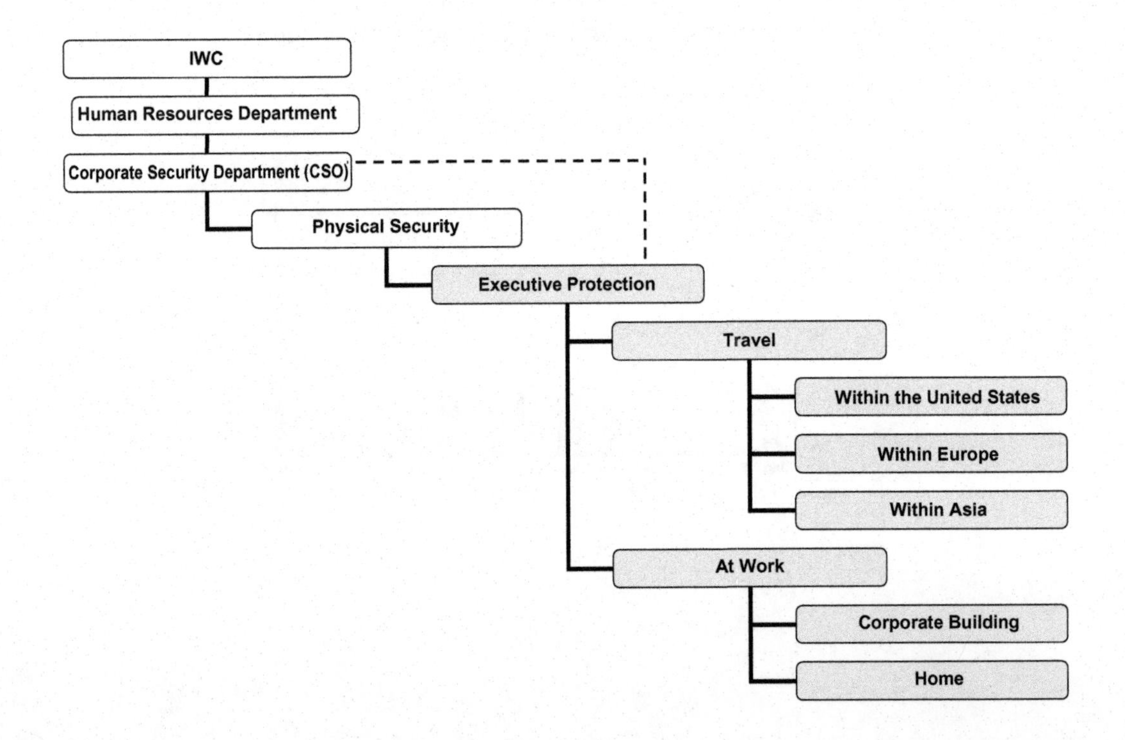
IWC
Human Resources Department
Corporate Security Department (CSO)
Physical Security
Executive Protection
Travel
Within the United States
Within Europe
Within Asia
At Work
Corporate Building
Home

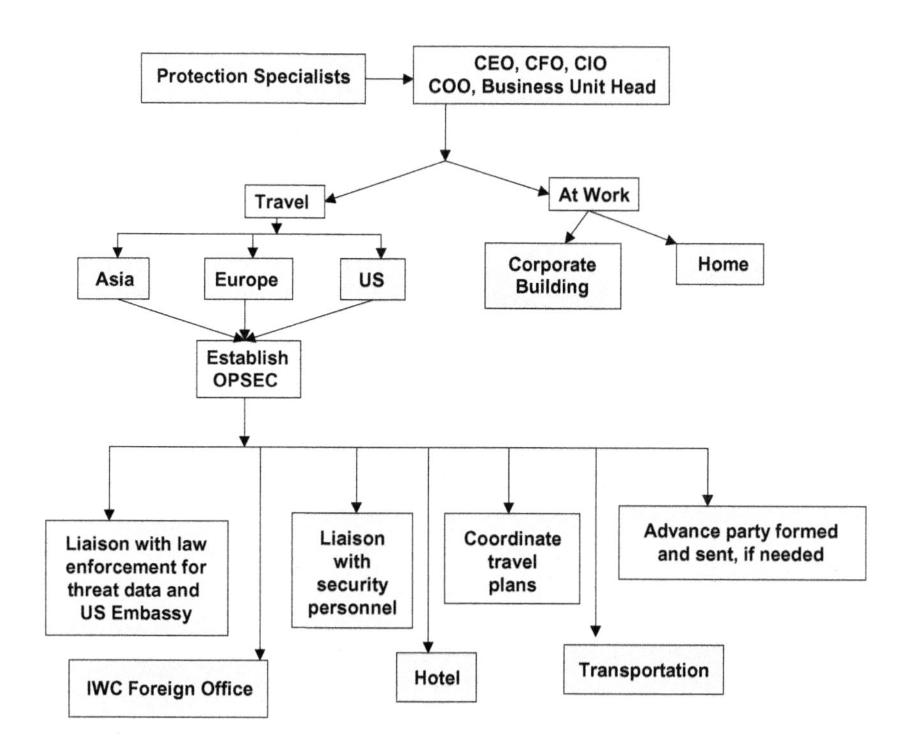

IWC Executive Protection Trips by Executives—2005

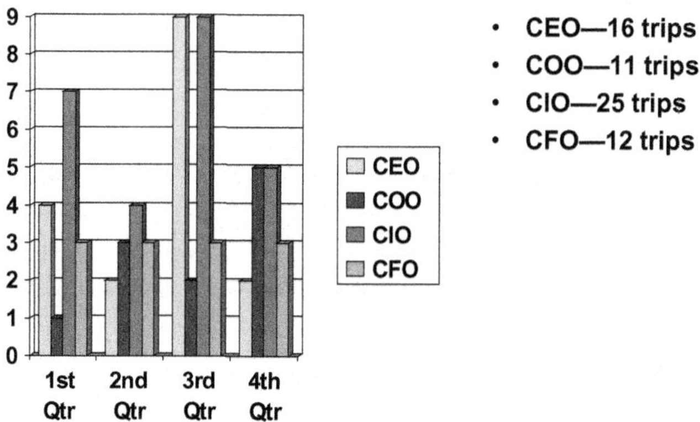

- CEO—16 trips
- COO—11 trips
- CIO—25 trips
- CFO—12 trips

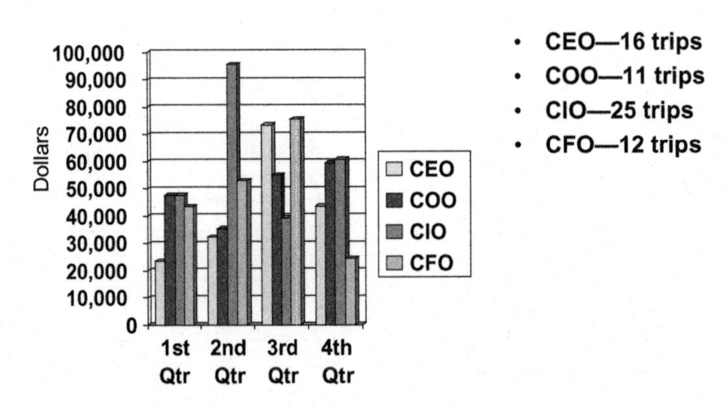
IWC Executive Protection Trips by Executives
Costs—2005
Dollars
100,000
90,000
80,000
70,000
60,000
50,000
40,000
30,000
20,000
10,000
0
1st Qtr
2nd Qtr
3rd Qtr
4th Qtr
CEO
COO
CIO
CFO
• CEO—16 trips
• COO—11 trips
• CIO—25 trips
• CFO—12 trips

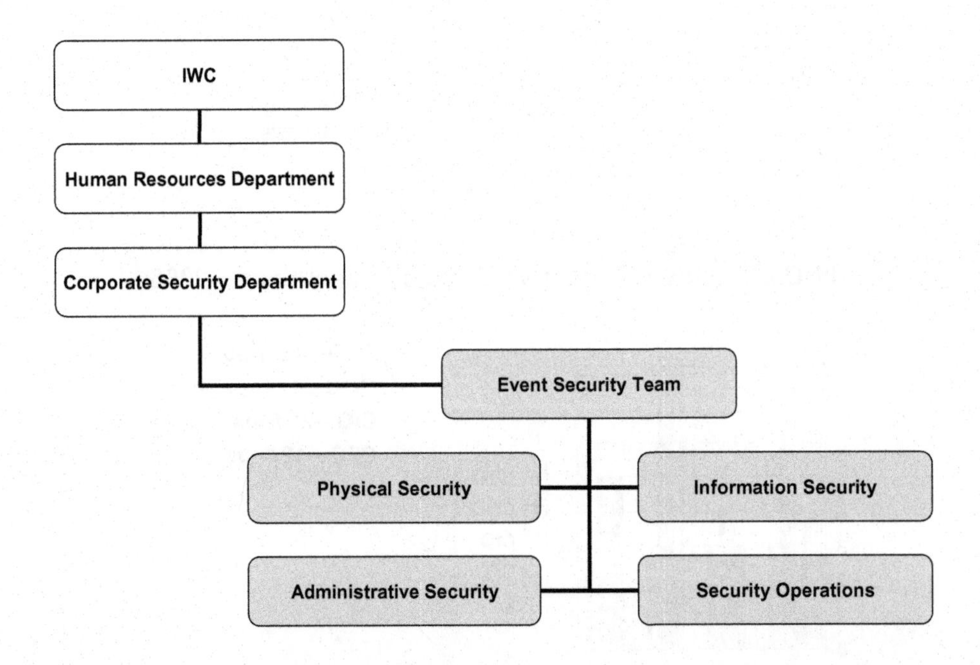
IWC
Human Resources Department
Corporate Security Department
Event Security Team
Physical Security
Information Security
Administrative Security
Security Operations

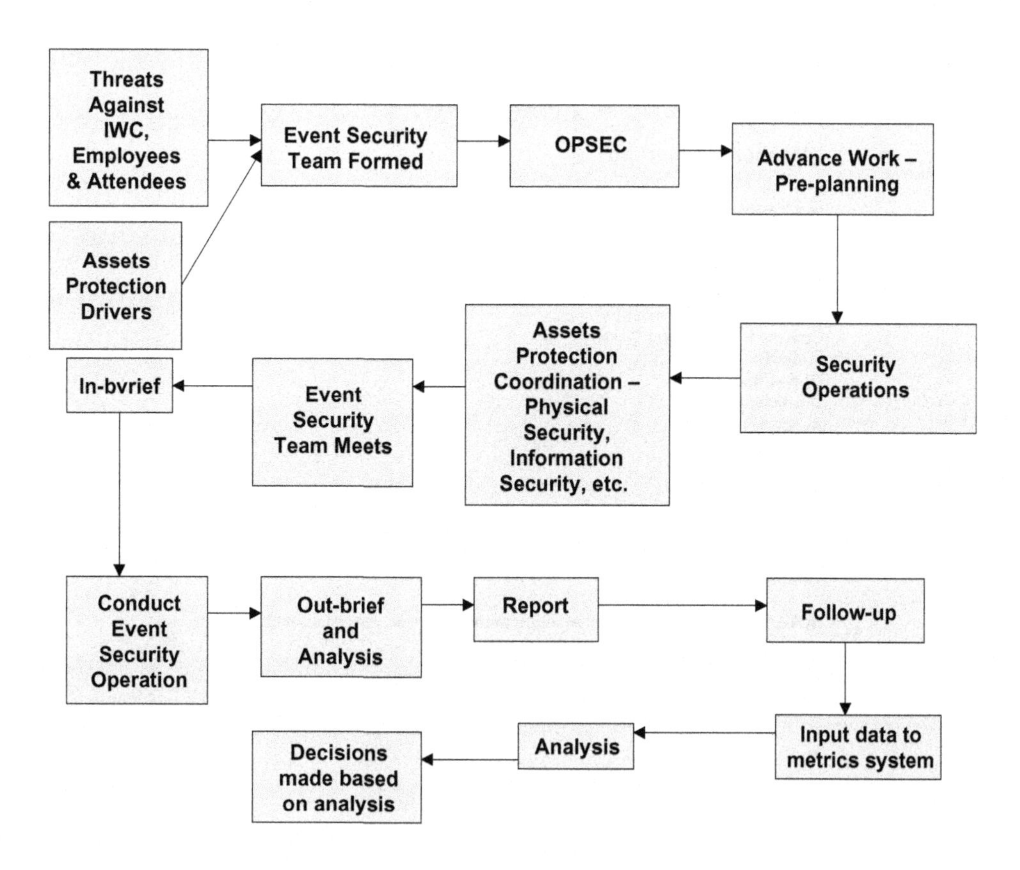
Threats Against IWC, Employees & Attendees
Assets Protection Drivers
Event Security Team Formed
OPSEC
Advance Work – Pre-planning
Security Operations
Assets Protection Coordination – Physical Security, Information Security, etc.
Event Security Team Meets
In-bvrief
Conduct Event Security Operation
Out-brief and Analysis
Report
Follow-up
Input data to metrics system
Analysis
Decisions made based on analysis

EVENT SECURITY BUDGET ANALYSES

Event	2004	2005
Stockholders Meeting		
ID Threats		
ID Risks		
ID Assets Involved		
ID Team and their costs		
Implement OPSEC		
Set Up Security Ops Center		
Operate Security Ops Center		
In-Brief		
Operation		
Out-Brief		
Analyses		
Report Written		
Report Published		
Report Distributed		
Follow-up Work		
Total		
Variance		

NOTE: Costs include personnel, equipment, supplies, records checks, etc.

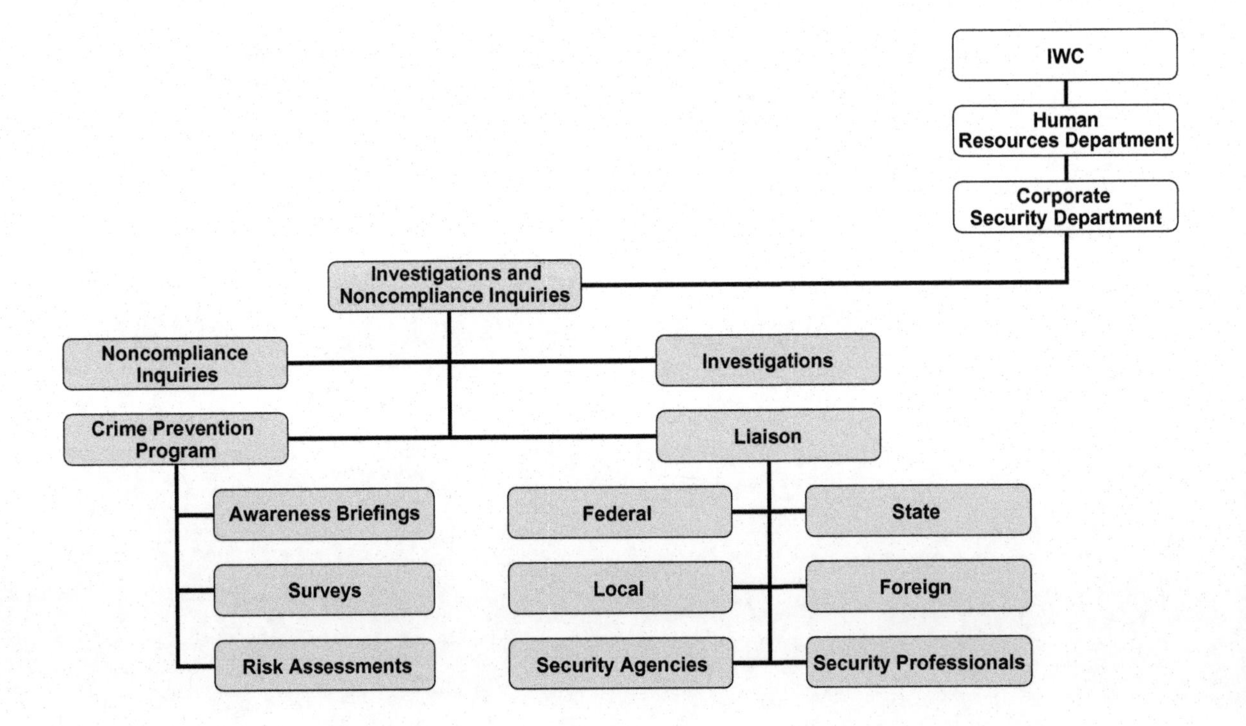

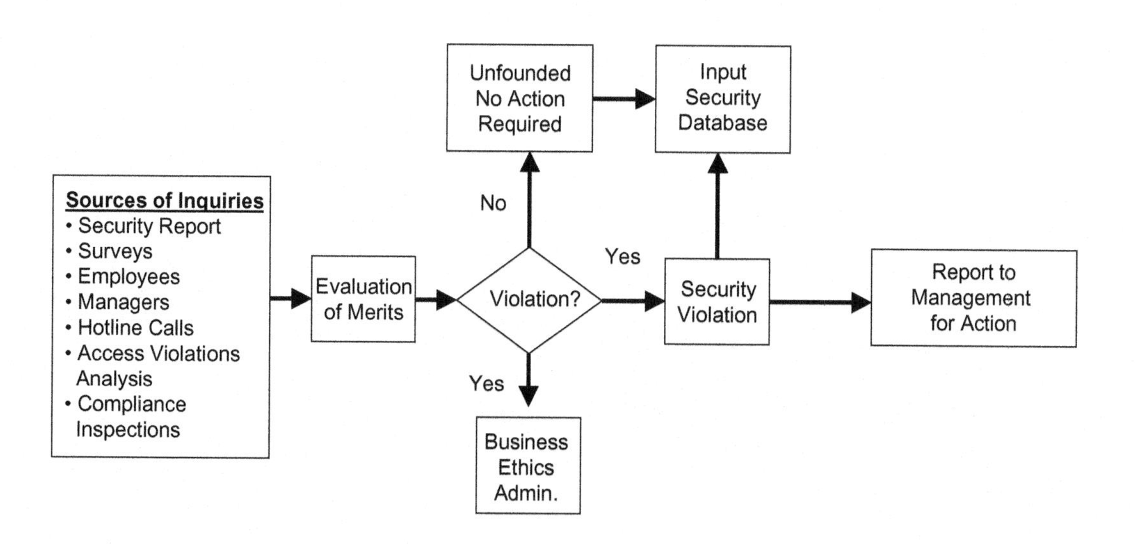
Unfounded No Action Required
Input Security Database
Sources of Inquiries
• Security Report
• Surveys
• Employees
• Managers
• Hotline Calls
• Access Violations Analysis
• Compliance Inspections
Evaluation of Merits
Violation?
No
Yes
Yes
Security Violation
Report to Management for Action
Business Ethics Admin.

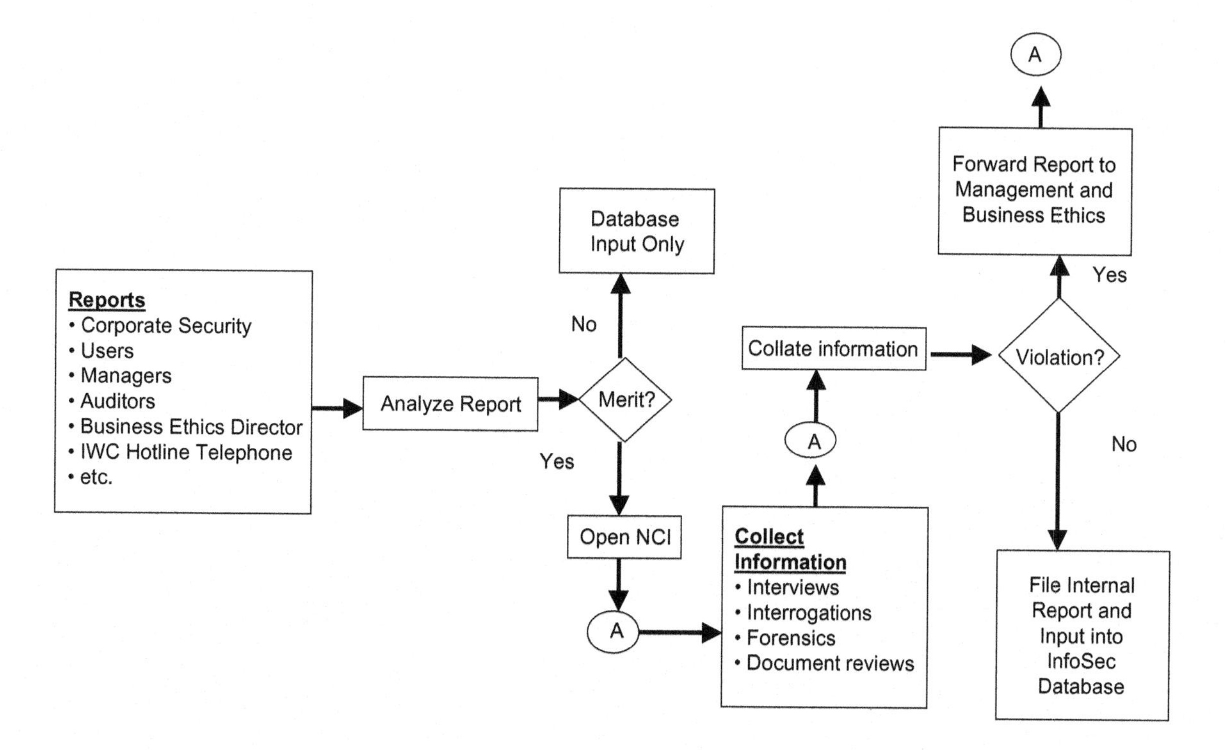
A
Forward Report to Management and Business Ethics
Database Input Only
Yes
Reports
• Corporate Security
• Users
• Managers
• Auditors
• Business Ethics Director
• IWC Hotline Telephone
• etc.
Analyze Report
No
Merit?
Yes
Collate information
Violation?
No
A
Open NCI
A
Collect Information
• Interviews
• Interrogations
• Forensics
• Document reviews
File Internal Report and Input into InfoSec Database

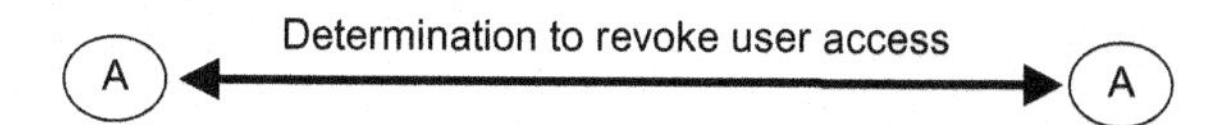
A
Determination to revoke user access
A

Number of Technology Related NCIs by Type—2005

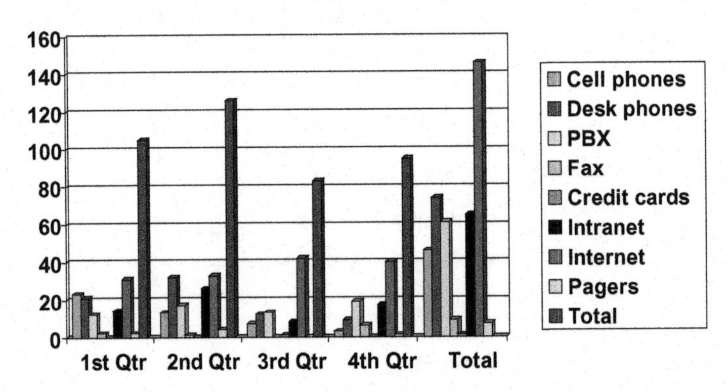

Noncompliance Inquiries—2005

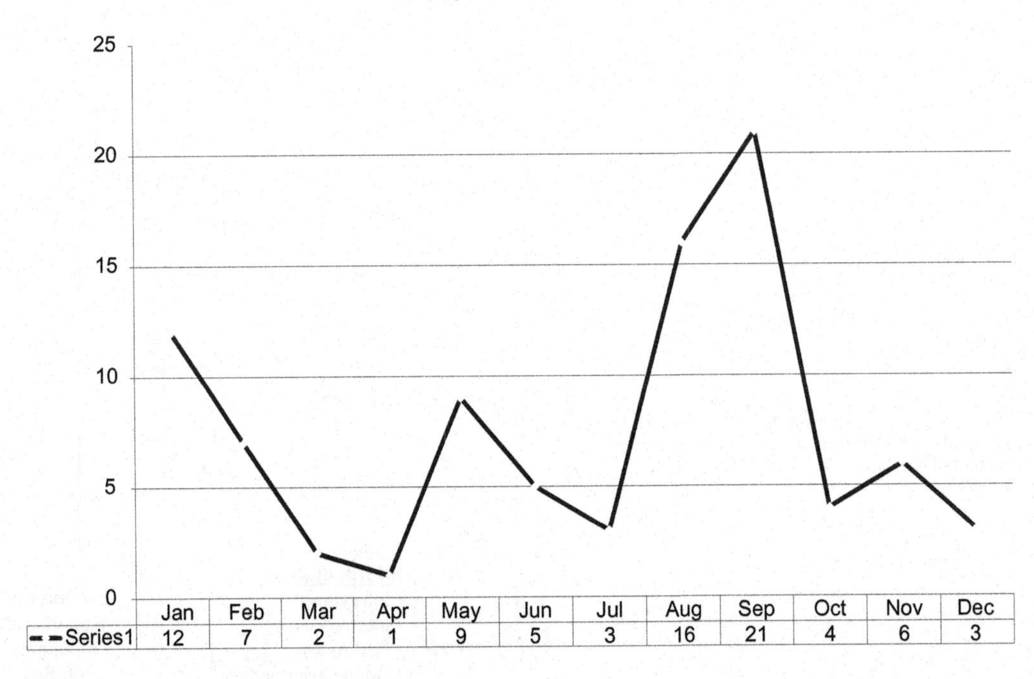

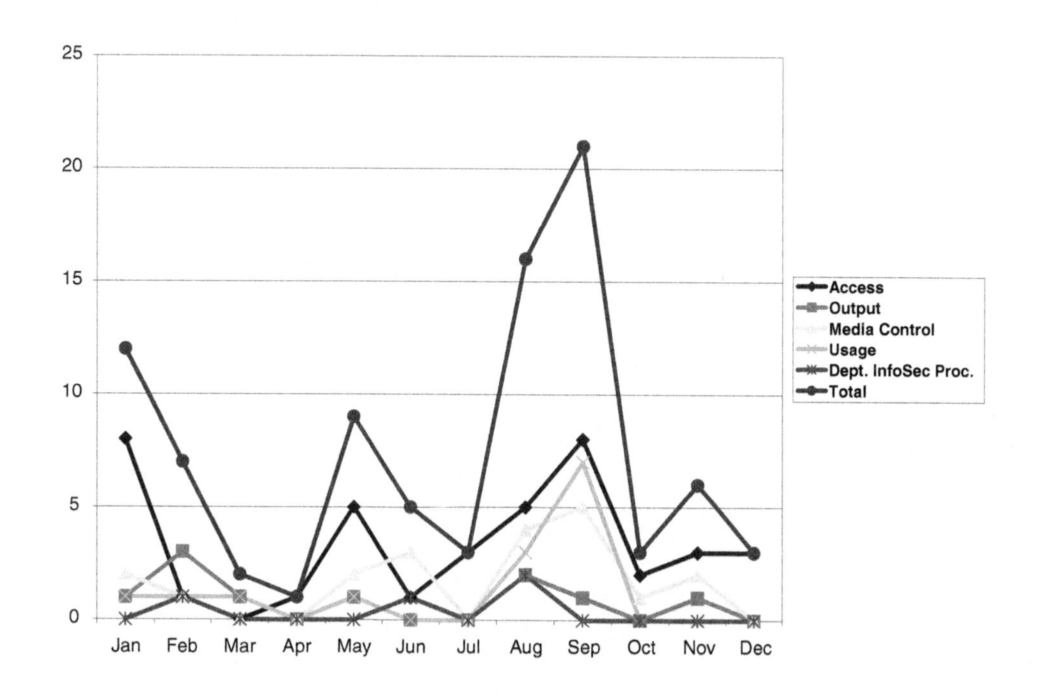
25
20
15
10
5
0
Jan
Feb
Mar
Apr
May
Jun
Jul
Aug
Sep
Oct
Nov
Dec
Access
Output
Media Control
Usage
Dept. InfoSec Proc.
Total

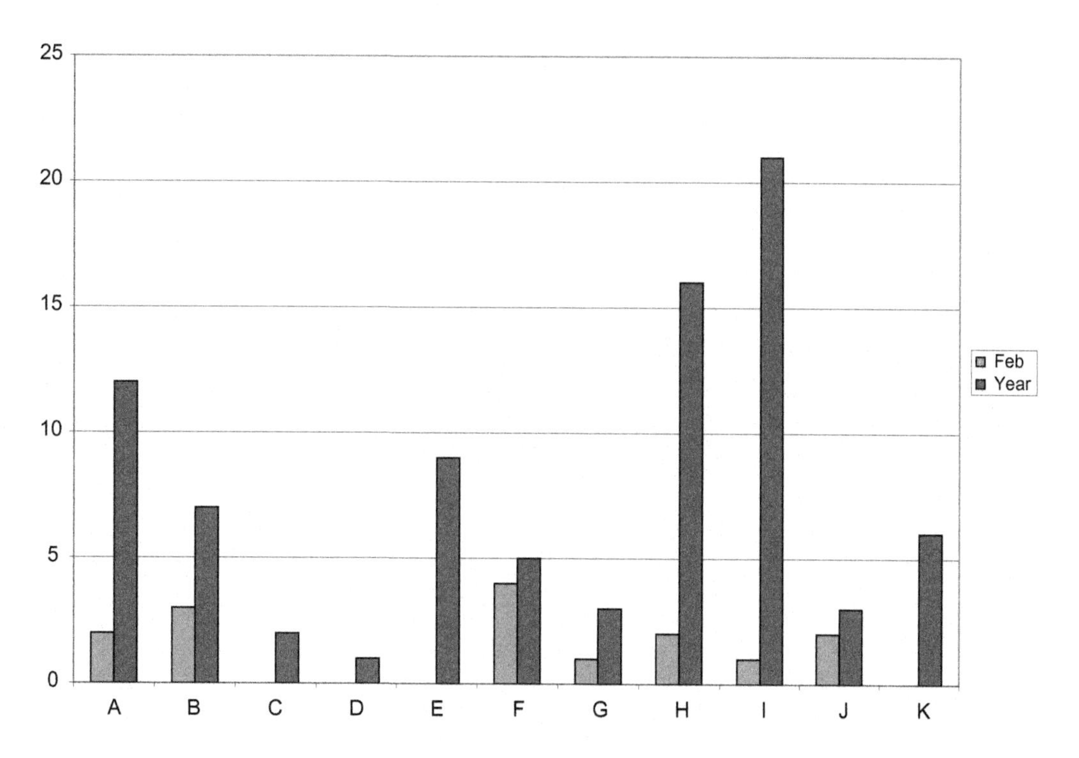
25
20
15
10
5
0
A
B
C
D
E
F
G
H
I
J
K
Feb
Year

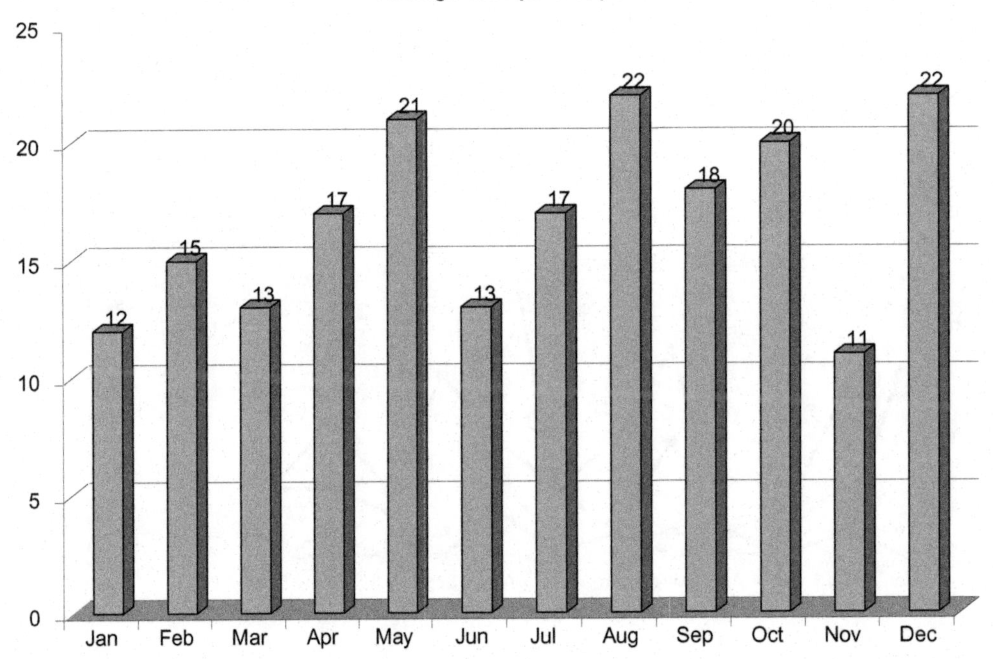
Average Time per NCI per Month—2005
25
20
15
10
5
0
12
15
13
17
21
13
17
22
18
20
11
22
Jan
Feb
Mar
Apr
May
Jun
Jul
Aug
Sep
Oct
Nov
Dec

Investigations and NCI Resource Allocation's Drivers

- The number of IWC employees has increased based on IWC's need to rapidly build up the workforce to handle the new contract work.
- The number of noncompliance inquiries has increased during that same time period.
- The number of investigations has increased during that same period of time.
- This increased workload has caused some delays in completing the inquiries and investigations in the 30-day period that was set as the goal.
- The ratio of incidents compared to the total number of employees indicates:
  - Personnel may not be getting sufficient information during their new-hire briefings.
  - The personnel being hired may not be thoroughly screened prior to hiring.

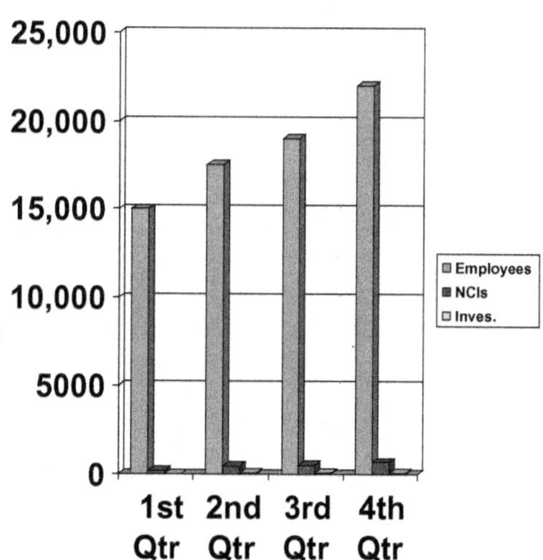

Number of New NCIs per Month—All Locations—2005

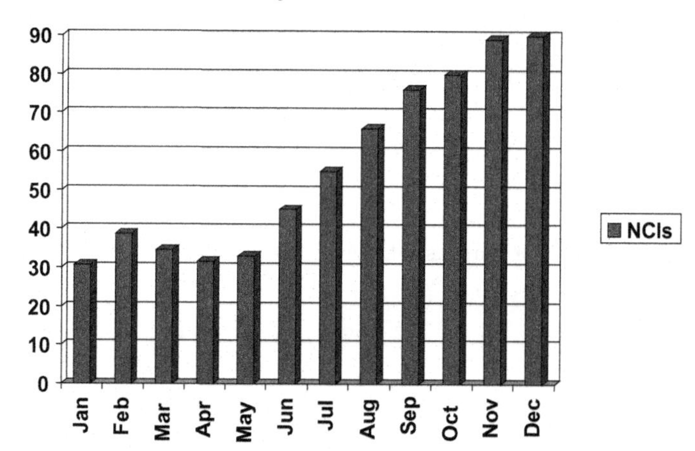

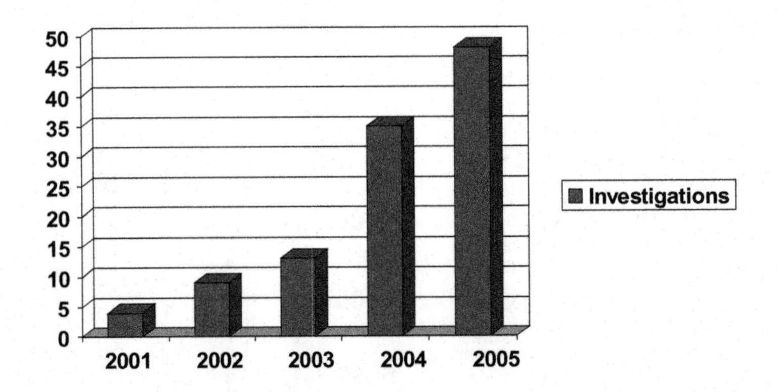
Number of Investigations—All Locations—2001–2005
50
45
40
35
30
25
20
15
10
5
0
2001
2002
2003
2004
2005
Investigations

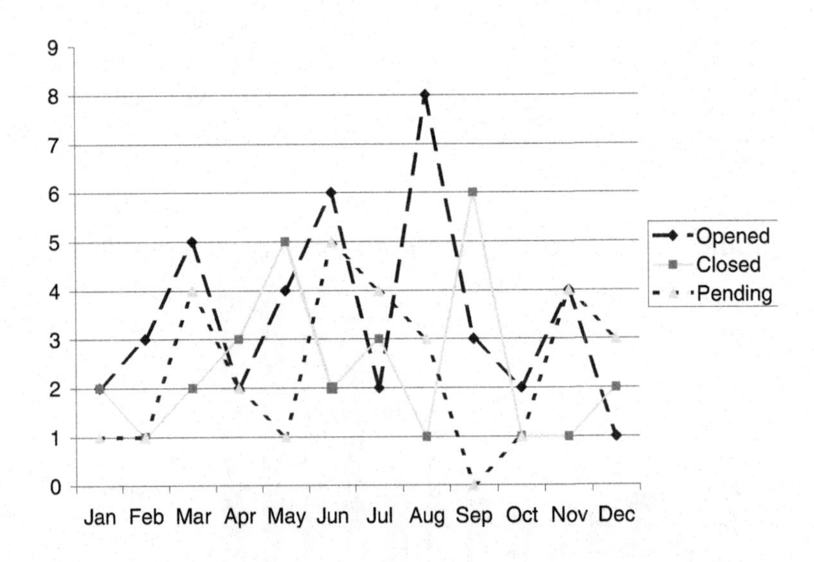
9
8
7
6
5
4
3
2
1
0
Jan Feb Mar Apr May Jun Jul Aug Sep Oct Nov Dec
Opened
Closed
Pending

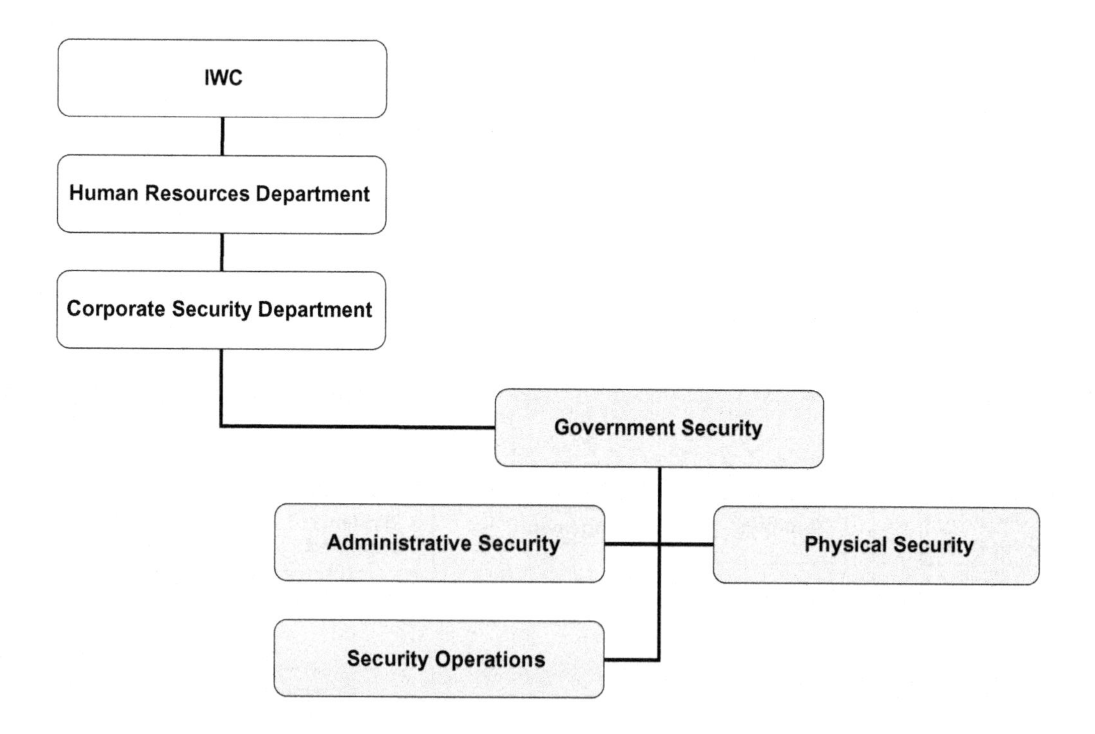
IWC
Human Resources Department
Corporate Security Department
Government Security
Administrative Security
Physical Security
Security Operations

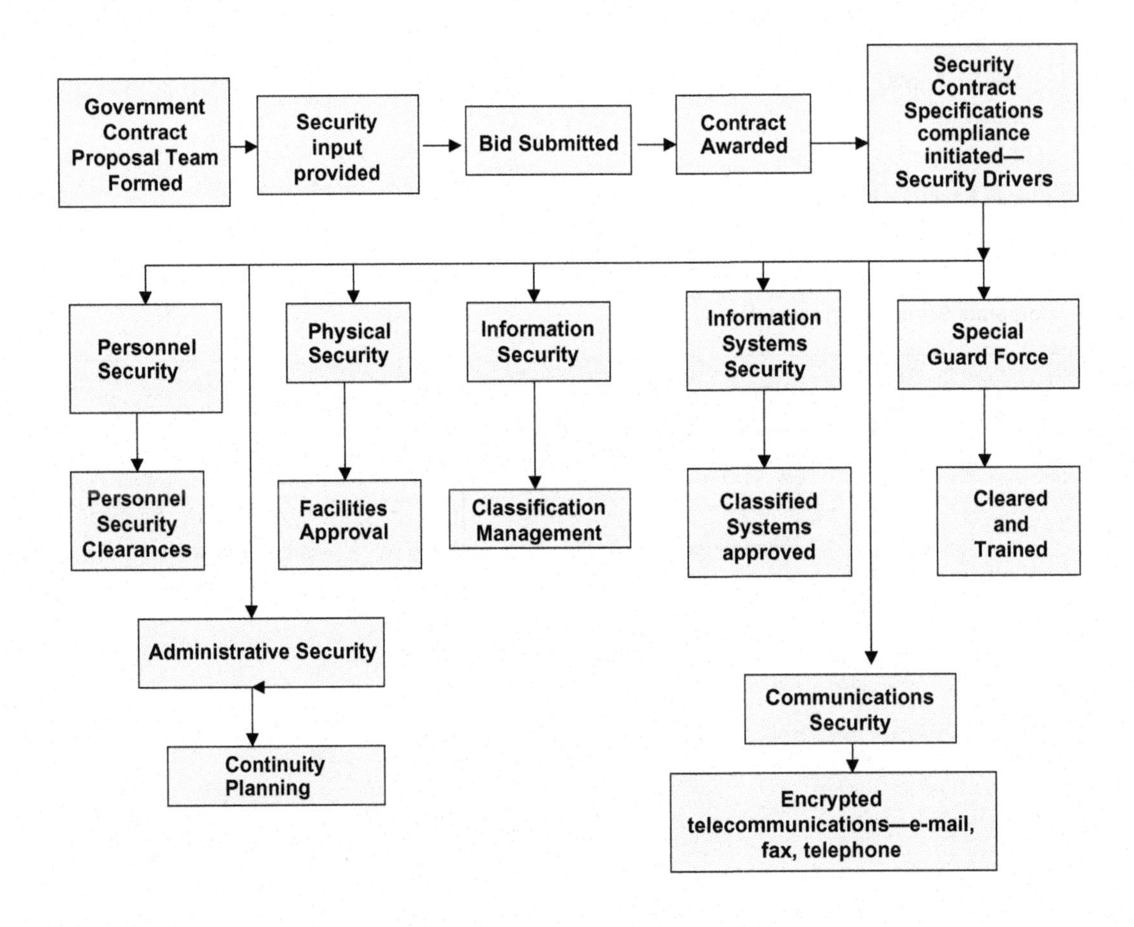

Facility Utilization Profile in Terms of Classified Activity

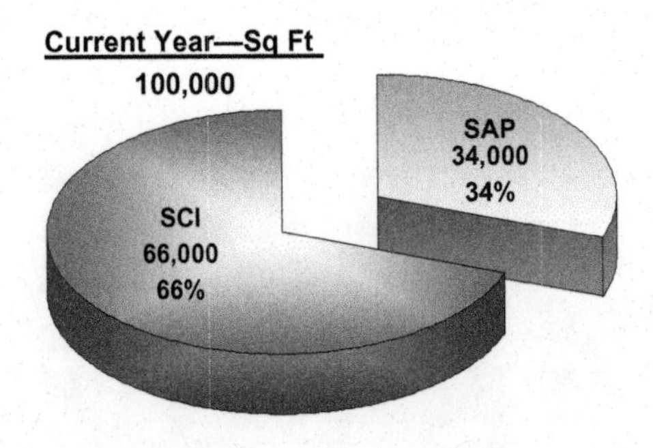

Employee Population and Security Clearances 1998–2005

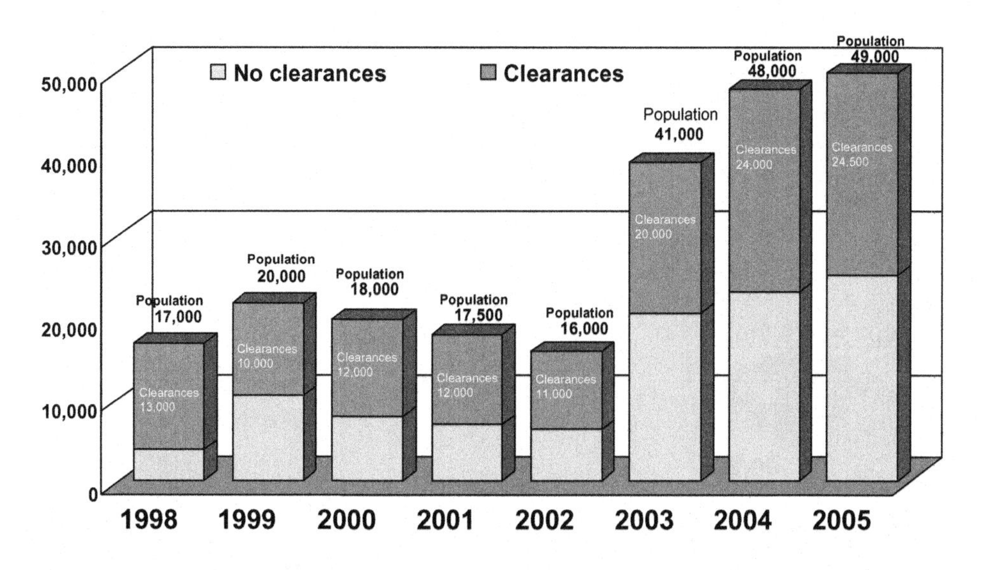

Special Accesses 1998–2005

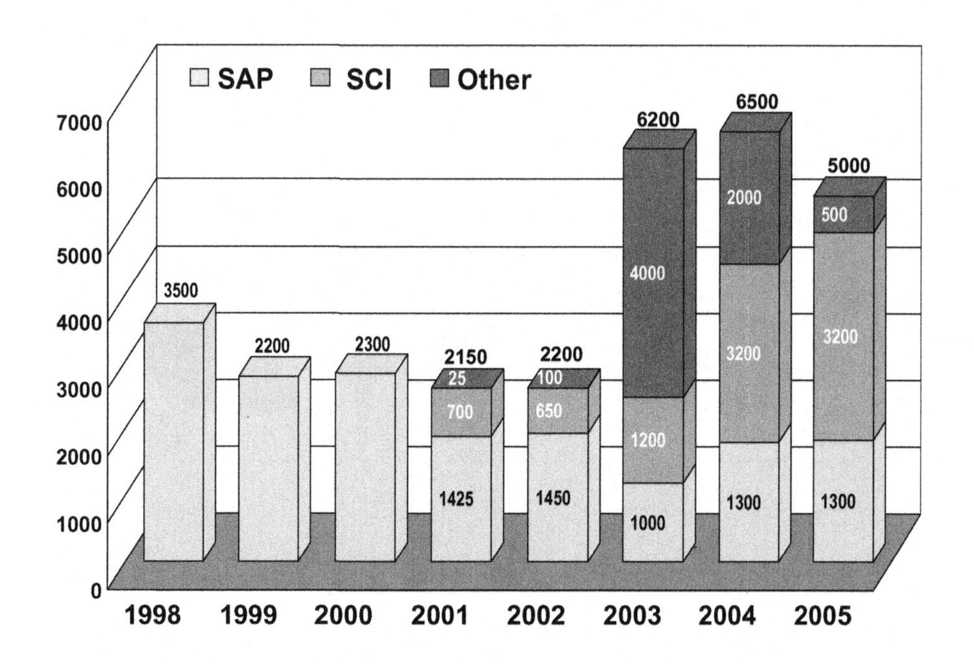

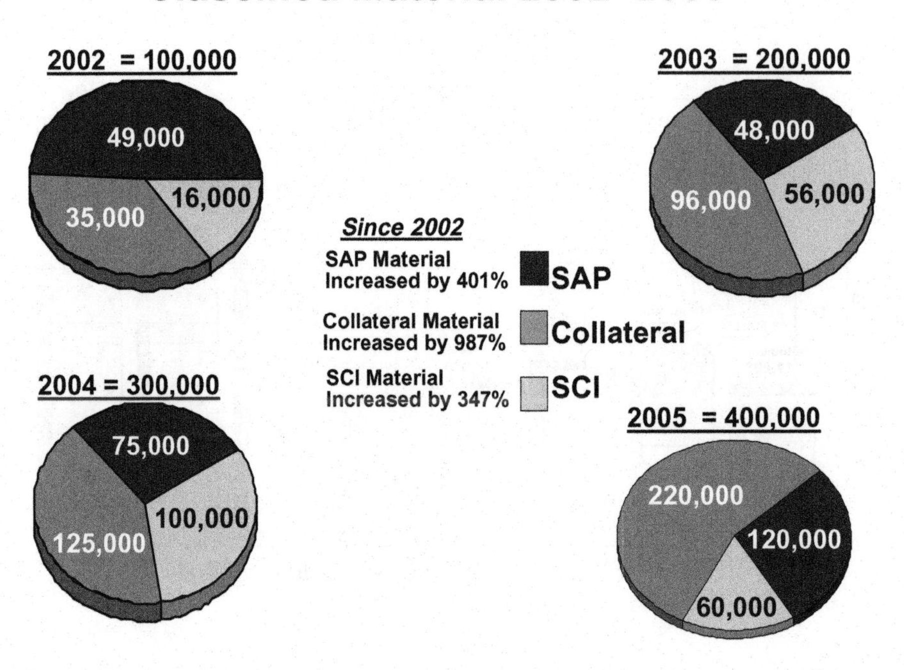
Classified Material 2002–2005
2002 = 100,000
49,000
35,000
16,000
2003 = 200,000
48,000
96,000
56,000
Since 2002
SAP Material
Increased by 401%
SAP
Collateral Material
Increased by 987%
Collateral
SCI Material
Increased by 347%
SCI
2004 = 300,000
75,000
125,000
100,000
2005 = 400,000
220,000
120,000
60,000

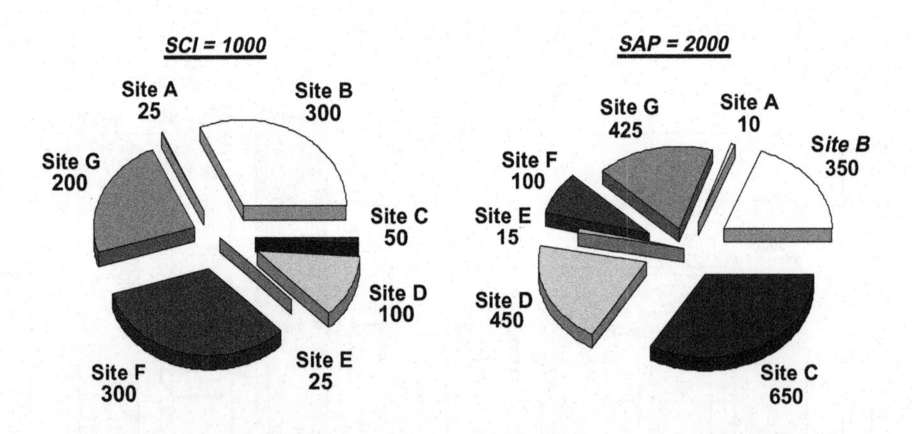
2005 Special Access by Site
SCI = 1000
Site A
25
Site B
300
Site G
200
Site C
50
Site D
100
Site E
25
Site F
300
SAP = 2000
Site G
425
Site A
10
Site B
350
Site F
100
Site E
15
Site D
450
Site C
650

DSS / Customers Inspections = 50
- Satisfactory—48
- Marginal—1
- Unsatisfactory—1
- Adverse reports submitted to DSS—30

Information Security
- Site classified LANs—29
- Site classified WANs—19
- Site classified stand-alones—350

Pre-Employment Investigations
PEIs performed—2500

Cases denoting derogatory information – 39
Total cost all PEI's—Approx. $250,000

Workplace Violence (WPV) Prevention Program
Incidents—12

Guard Service Outsourced		
	Use	Provider
Site A	Yes	Service provider A
Site B	Yes	Service provider A
Site C	Yes	Service provider A
Site D	No	Proprietary
Site E	Yes	Service provider A
Site F	Yes	Service provider A
Site G	Yes	Service provider A
Site H	No	Proprietary
Site I	Yes	Service provider A

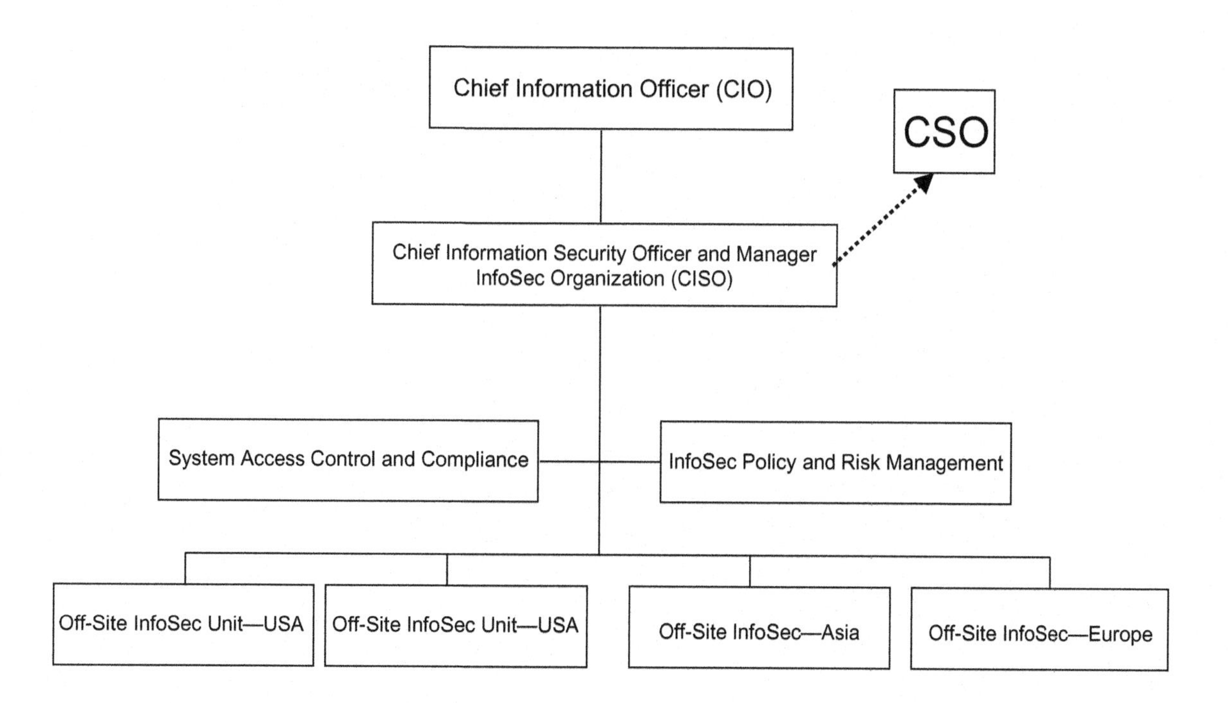

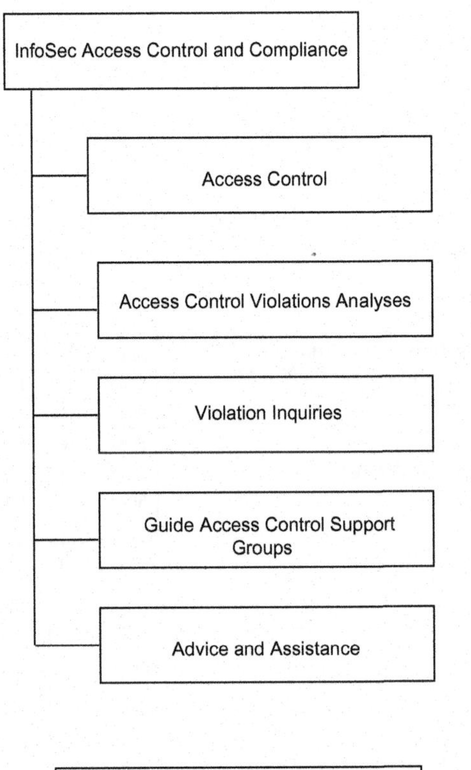
InfoSec Access Control and Compliance
Access Control
Access Control Violations Analyses
Violation Inquiries
Guide Access Control Support Groups
Advice and Assistance

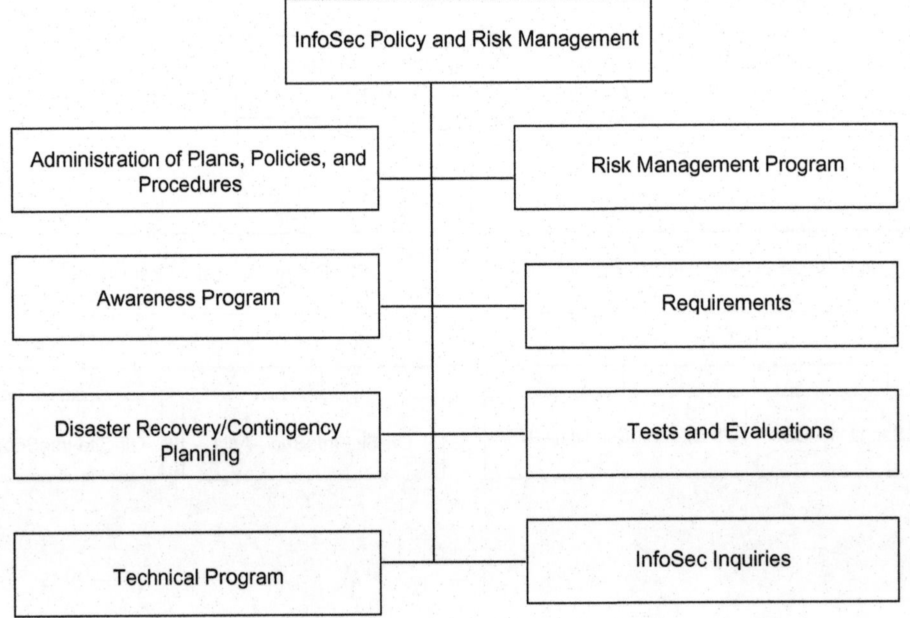
InfoSec Policy and Risk Management
Administration of Plans, Policies, and Procedures
Risk Management Program
Awareness Program
Requirements
Disaster Recovery/Contingency Planning
Tests and Evaluations
Technical Program
InfoSec Inquiries

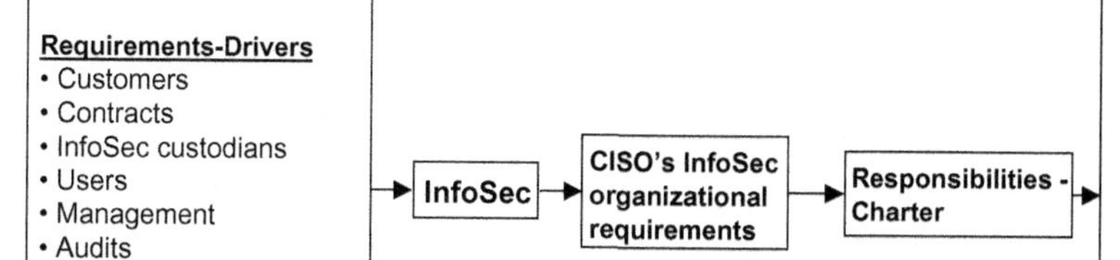
Requirements-Drivers
• Customers
• Contracts
• InfoSec custodians
• Users
• Management
• Audits
• Test and evaluations
• Other employees
• Laws
• Regulations
• Noncompliance inquiries
• Investigations
• IWC business plans
• CISO's plans
• Best business practices
• Best InfoSec practices
InfoSec
CISO's InfoSec organizational requirements
Responsibilities - Charter
CISO Organizational Functions
• Identification of InfoSec requirements
• Access controls
• Noncompliance inquiries (NCI)
• Disaster recovery/emergency planning
• Tests and evaluations
• Intranet security
• Internet and web site security
• Security applications protection
• Security software development
• Software interface InfoSec evaluations
• Access control violations analyses
• Aystems' approvals
• InfoSec awareness and training
• Contractual compliance inspections
• InfoSec risk management

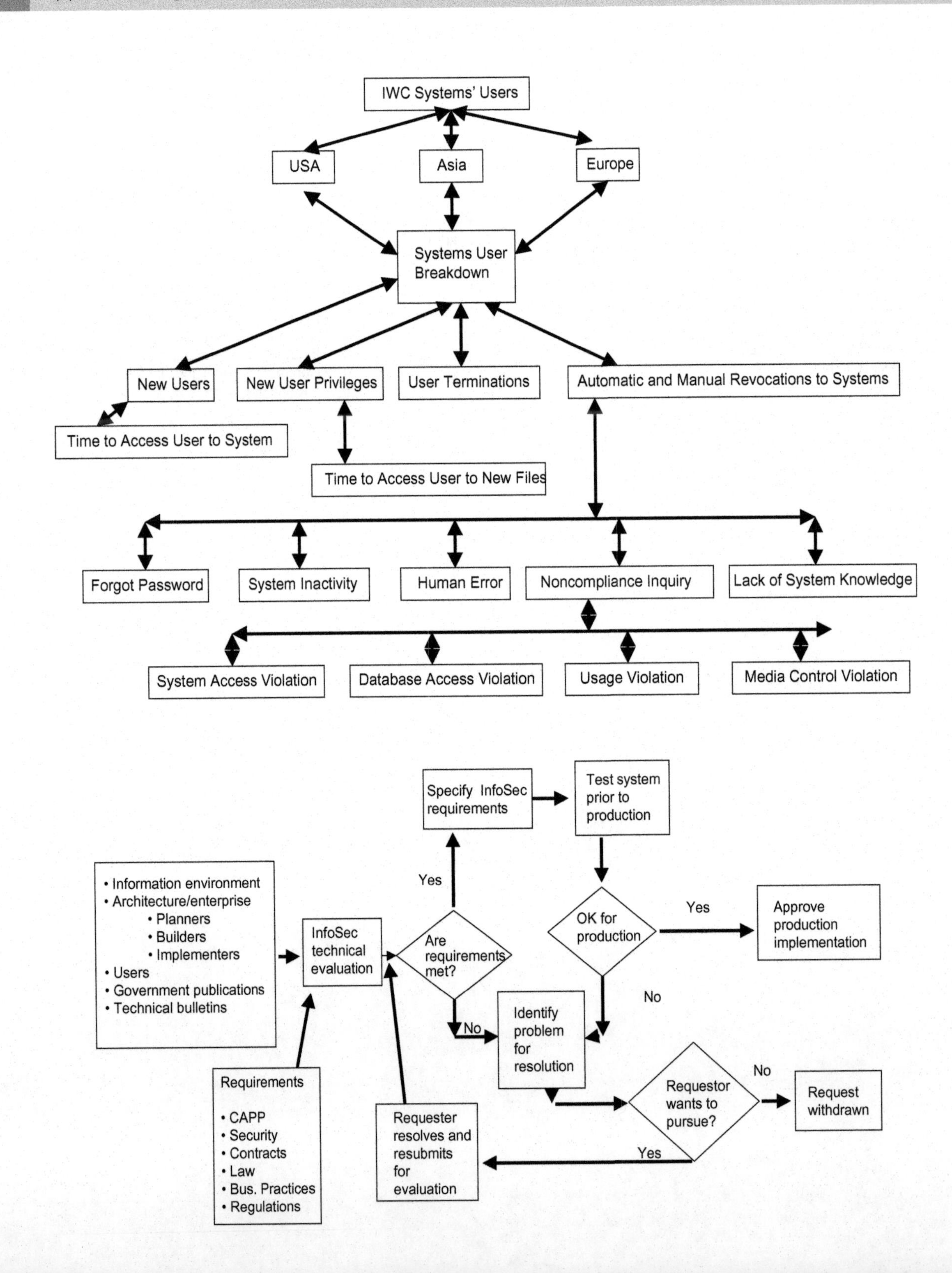

IWC Systems' Users
USA
Asia
Europe
Systems User Breakdown
New Users
New User Privileges
User Terminations
Automatic and Manual Revocations to Systems
Time to Access User to System
Time to Access User to New Files
Forgot Password
System Inactivity
Human Error
Noncompliance Inquiry
Lack of System Knowledge
System Access Violation
Database Access Violation
Usage Violation
Media Control Violation
Specify InfoSec requirements
Test system prior to production
• Information environment
• Architecture/enterprise
• Planners
• Builders
• Implementers
• Users
• Government publications
• Technical bulletins
InfoSec technical evaluation
Yes
Are requirements met?
No
OK for production
Yes
Approve production implementation
No
Identify problem for resolution
Requirements
• CAPP
• Security
• Contracts
• Law
• Bus. Practices
• Regulations
Requester resolves and resubmits for evaluation
Requestor wants to pursue?
No
Request withdrawn
Yes

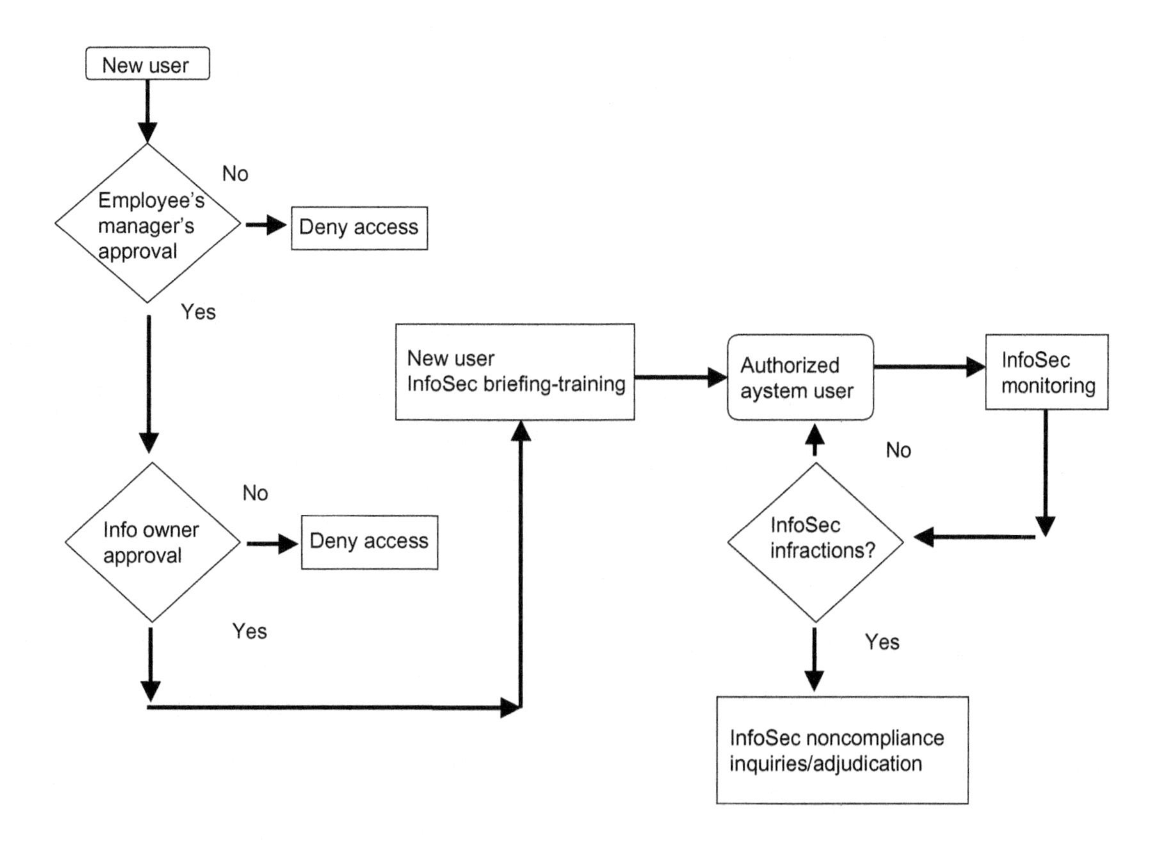
New user
Employee's manager's approval
No
Deny access
Yes
Info owner approval
No
Deny access
Yes
New user InfoSec briefing-training
Authorized aystem user
InfoSec monitoring
No
InfoSec infractions?
Yes
InfoSec noncompliance inquiries/adjudication

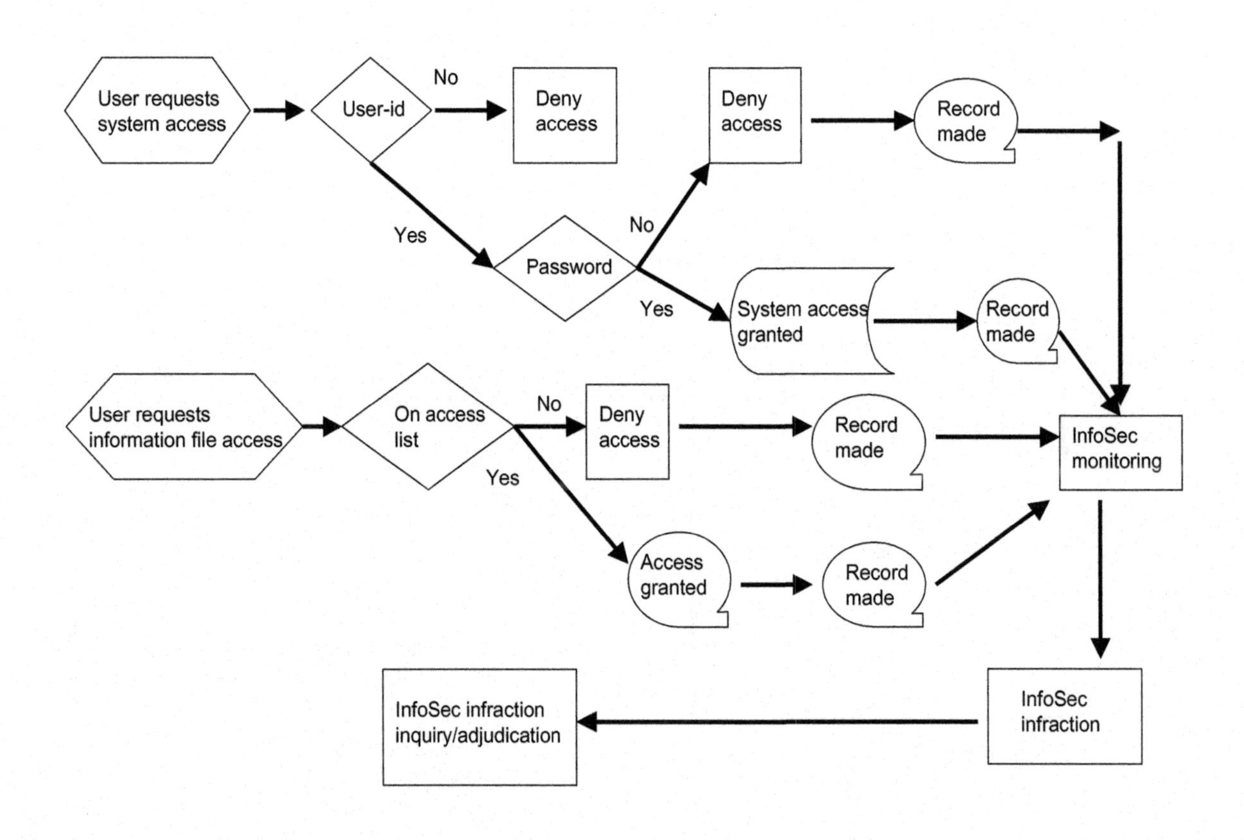
User requests system access
User-id
No
Deny access
Yes
Password
No
Deny access
Record made
Yes
System access granted
Record made
User requests information file access
On access list
No
Deny access
Record made
InfoSec monitoring
Yes
Access granted
Record made
InfoSec infraction
InfoSec infraction inquiry/adjudication

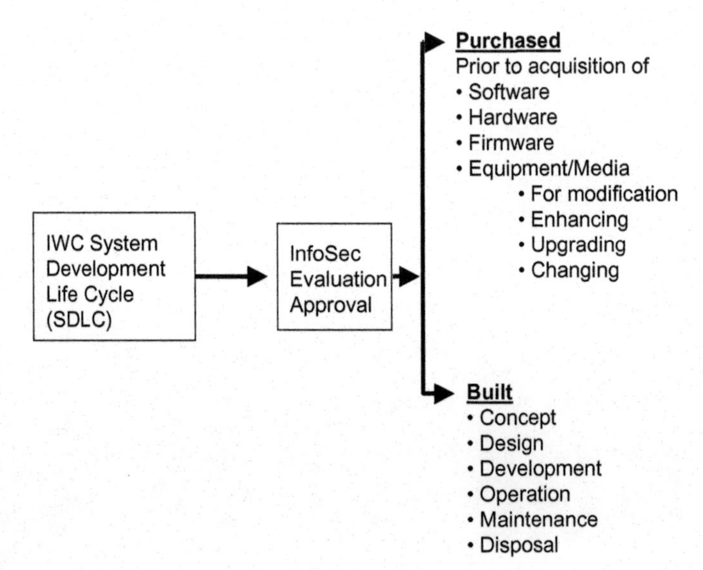
IWC System Development Life Cycle (SDLC)
InfoSec Evaluation Approval
Purchased
Prior to acquisition of
• Software
• Hardware
• Firmware
• Equipment/Media
• For modification
• Enhancing
• Upgrading
• Changing
Built
• Concept
• Design
• Development
• Operation
• Maintenance
• Disposal

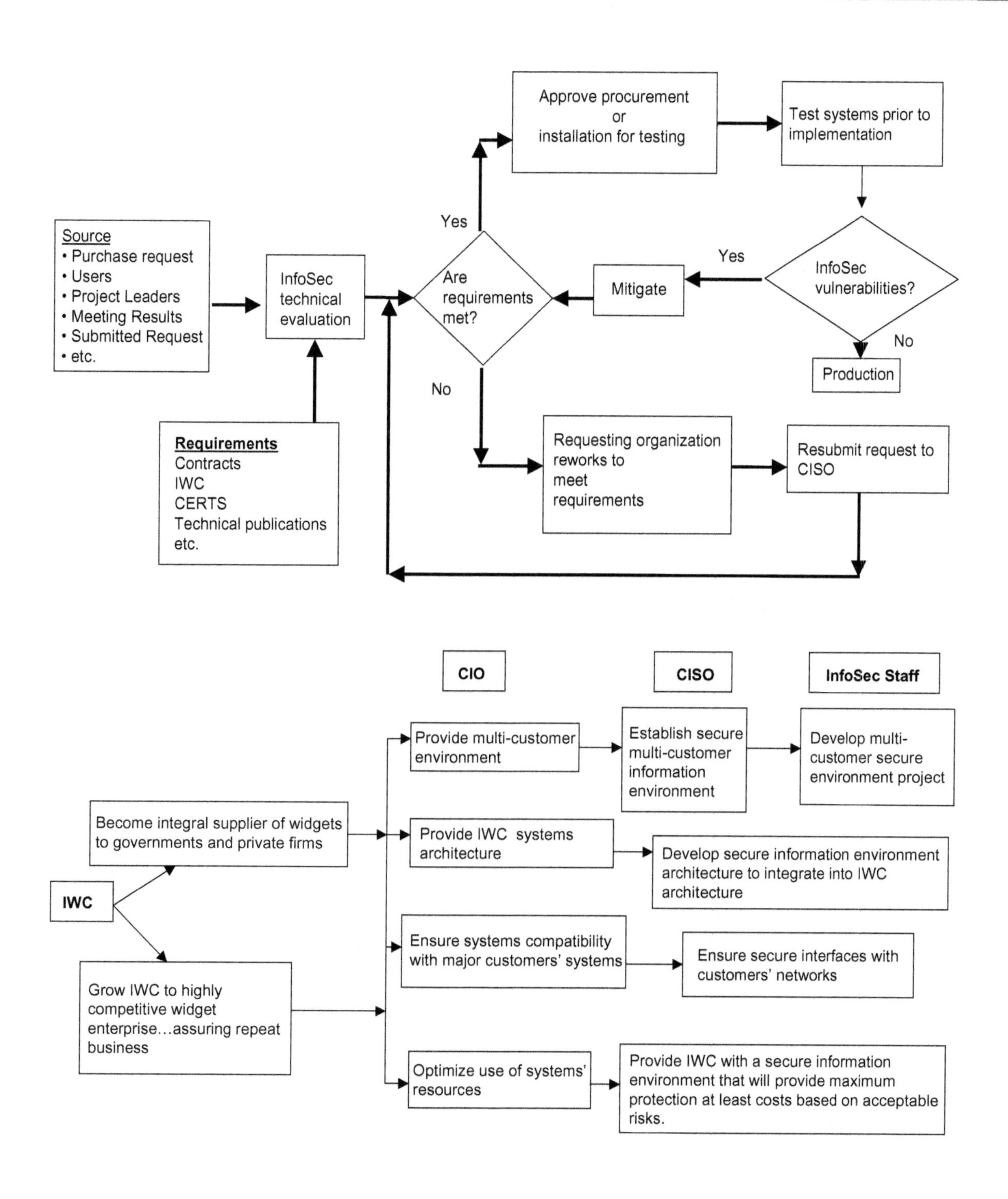
Approve procurement
or
installation for testing
Test systems prior to implementation
Yes
Source
• Purchase request
• Users
• Project Leaders
• Meeting Results
• Submitted Request
• etc.
InfoSec technical evaluation
Are requirements met?
Mitigate
Yes
InfoSec vulnerabilities?
No
Production
No
Requirements
Contracts
IWC
CERTS
Technical publications
etc.
Requesting organization reworks to meet requirements
Resubmit request to CISO
CIO
CISO
InfoSec Staff
Provide multi-customer environment
Establish secure multi-customer information environment
Develop multi-customer secure environment project
Become integral supplier of widgets to governments and private firms
Provide IWC systems architecture
Develop secure information environment architecture to integrate into IWC architecture
IWC
Ensure systems compatibility with major customers' systems
Ensure secure interfaces with customers' networks
Grow IWC to highly competitive widget enterprise…assuring repeat business
Optimize use of systems' resources
Provide IWC with a secure information environment that will provide maximum protection at least costs based on acceptable risks.

Total Users of IWC Networks World wide—2005

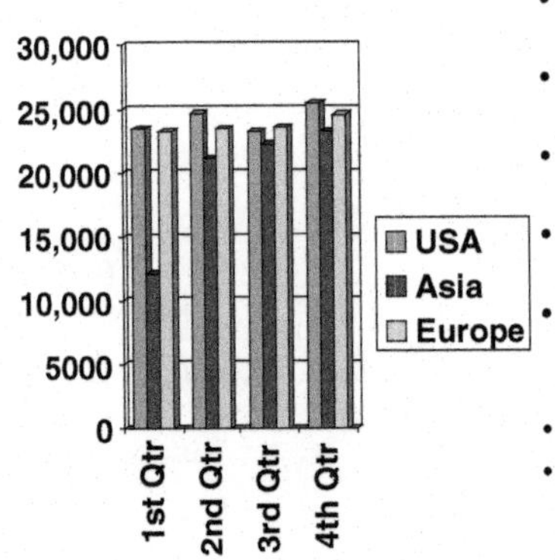

- Number of users **increasing** at all locations
- Average **increase** in users per quarter is **11%**
- Number of access violations has **increased** at a rate of **13%**
- Number of awareness briefings has **increased** at a rate of **10%**
- Number of noncompliance inquiries has **increased** at a rate of **17%**
- **No increase in InfoSec staff**
- **InfoSec functional productivity improvement kept headcount steady while handling increased workload without a decline in service quality**

Total Number of IWC Systems World wide—2005

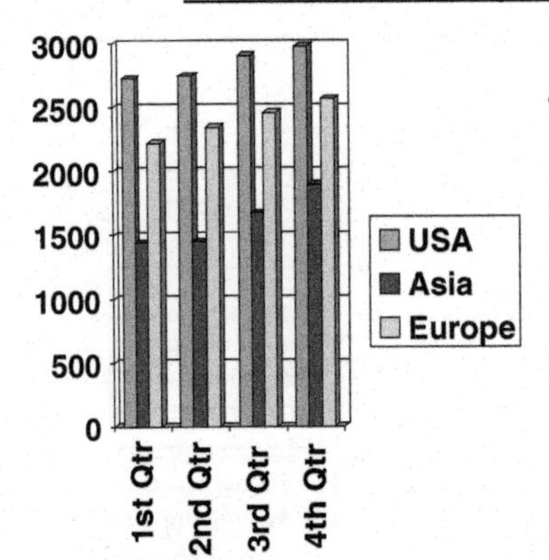

- Number of systems **increasing** at all locations at average of **8.5%**
- InfoSec workload **increasing** at a rate of **2.3%**
- Noncompliance Inquiries **increasing** at a rate of **.17%**
- **InfoSec manpower remains the same due to productivity gains with no decrease in quality of service**

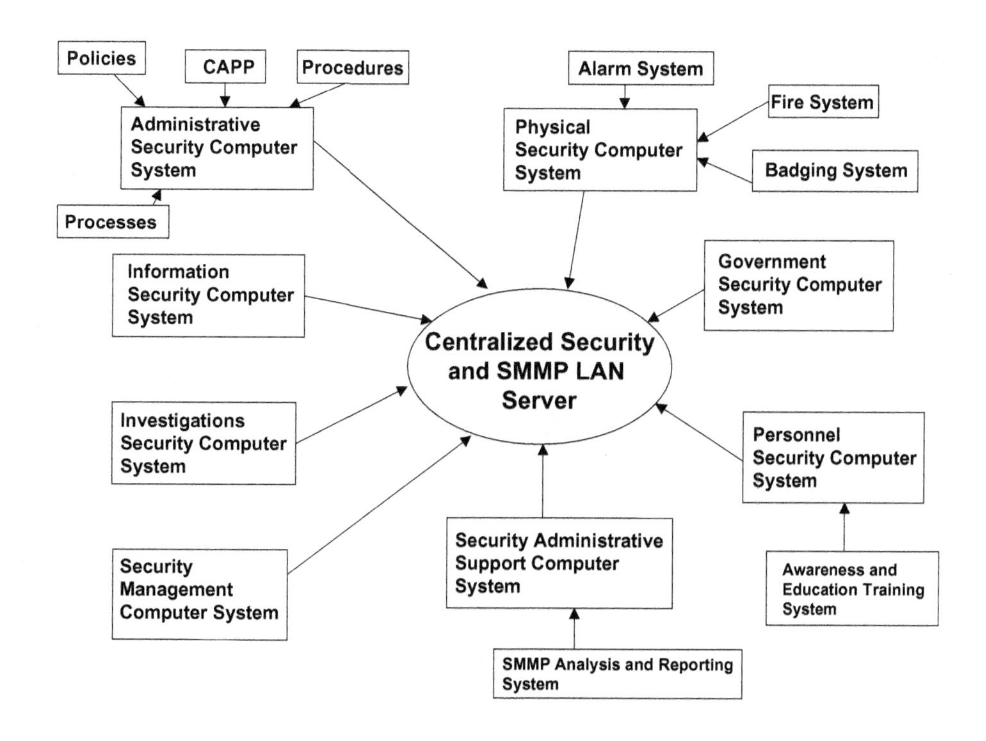
Policies
CAPP
Procedures
Administrative Security Computer System
Processes
Alarm System
Fire System
Physical Security Computer System
Badging System
Information Security Computer System
Government Security Computer System
Centralized Security and SMMP LAN Server
Investigations Security Computer System
Personnel Security Computer System
Security Management Computer System
Security Administrative Support Computer System
Awareness and Education Training System
SMMP Analysis and Reporting System

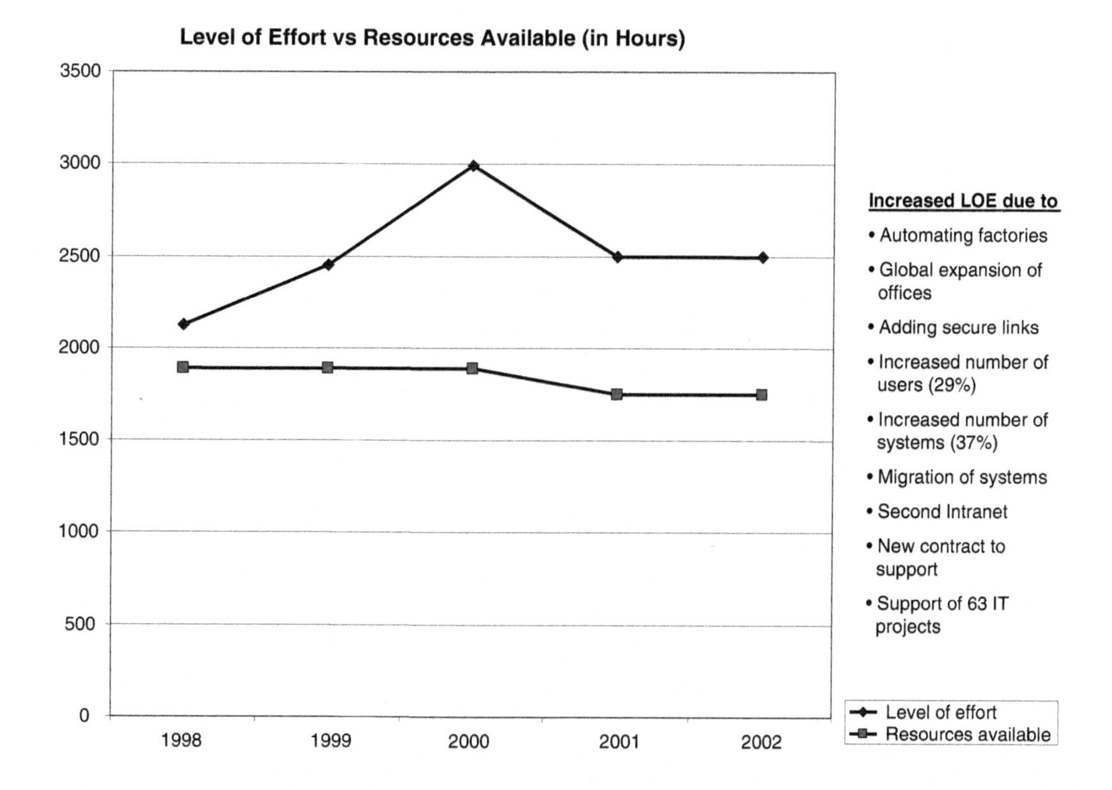
Level of Effort vs Resources Available (in Hours)
3500
3000
2500
2000
1500
1000
500
0
1998
1999
2000
2001
2002
Increased LOE due to
• Automating factories
• Global expansion of offices
• Adding secure links
• Increased number of users (29%)
• Increased number of systems (37%)
• Migration of systems
• Second Intranet
• New contract to support
• Support of 63 IT projects
Level of effort
Resources available

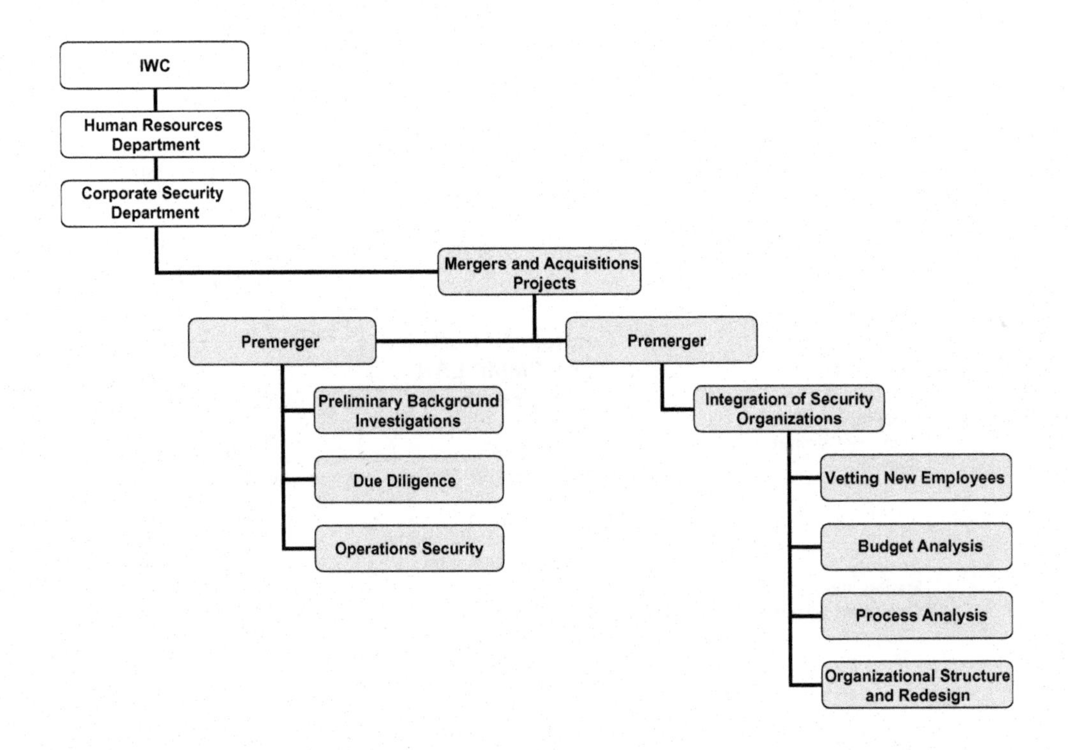

Merger and Acquisition Team Supporting a Large Transaction

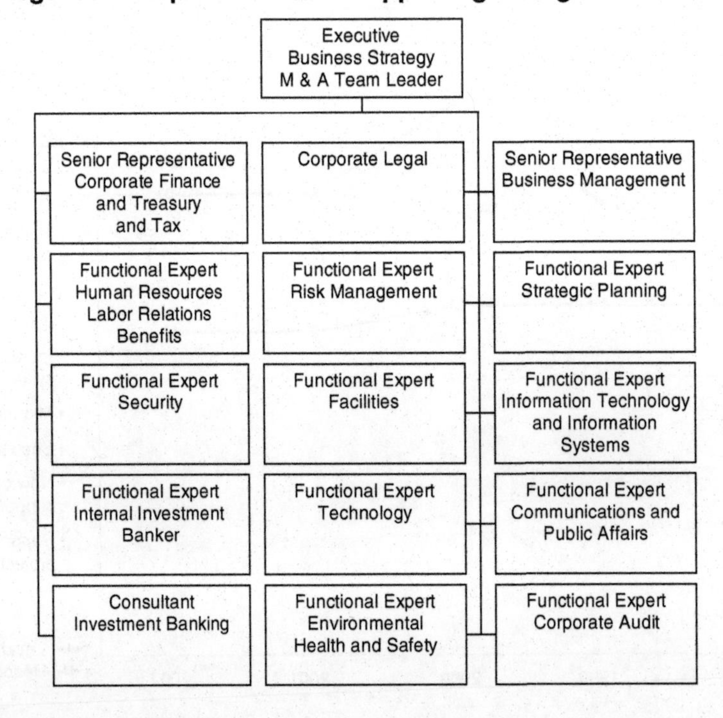

Postmerger and Acquisition Integration Team Membership Supporting a Large Integration

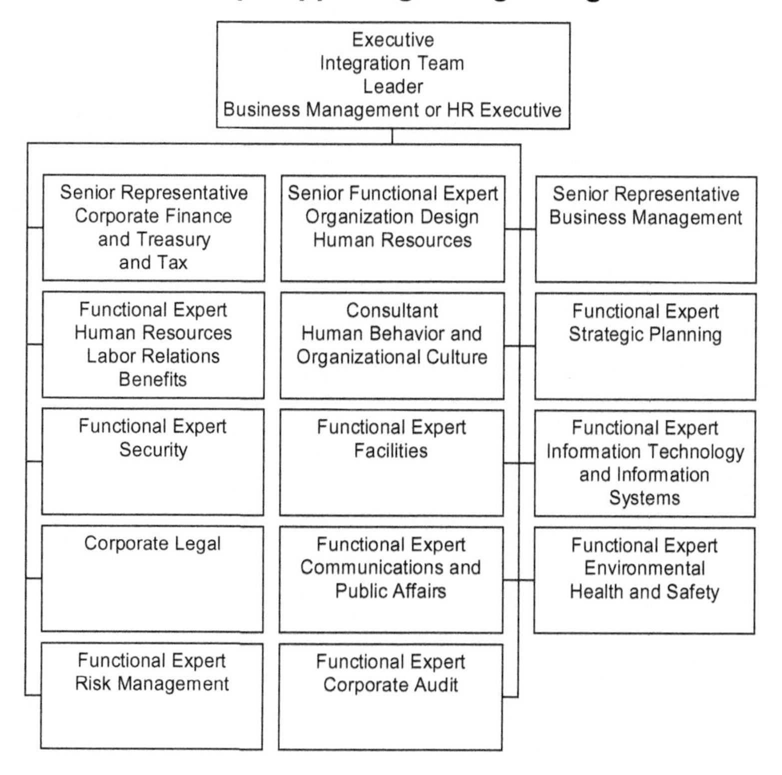

The Competitive Intelligence Collection Cycle

1. **Planning and Direction** - Determine what the business needs to know

2. **Collections** - Obtain the information needed

3. **Analyses** - Understanding what the information means

4. **Dissemination** - Getting the right information to the right people at the right time; obtain feedback; get re-tasked

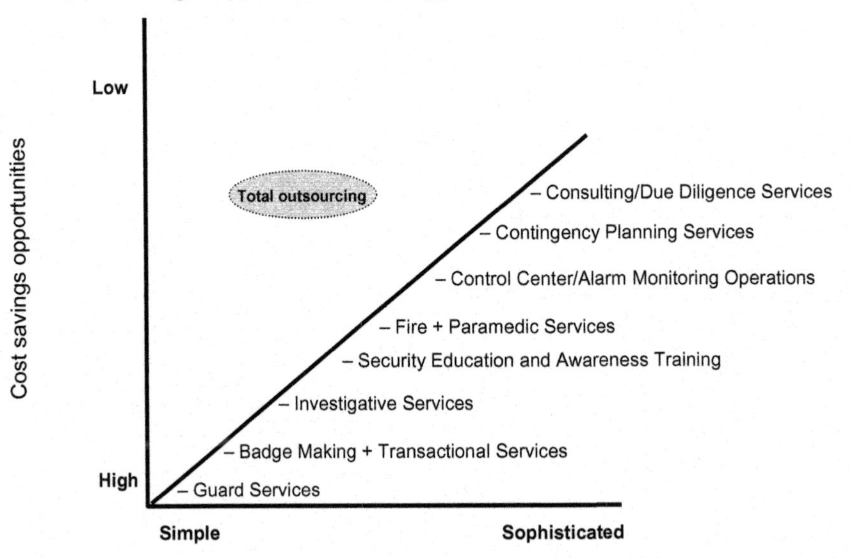
Cost Savings Opportunities vs Type of Process/Services Outsourced
Low
High
Cost savings opportunities
Total outsourcing
– Consulting/Due Diligence Services
– Contingency Planning Services
– Control Center/Alarm Monitoring Operations
– Fire + Paramedic Services
– Security Education and Awareness Training
– Investigative Services
– Badge Making + Transactional Services
– Guard Services
Simple
Sophisticated
Type of process/services outsourced

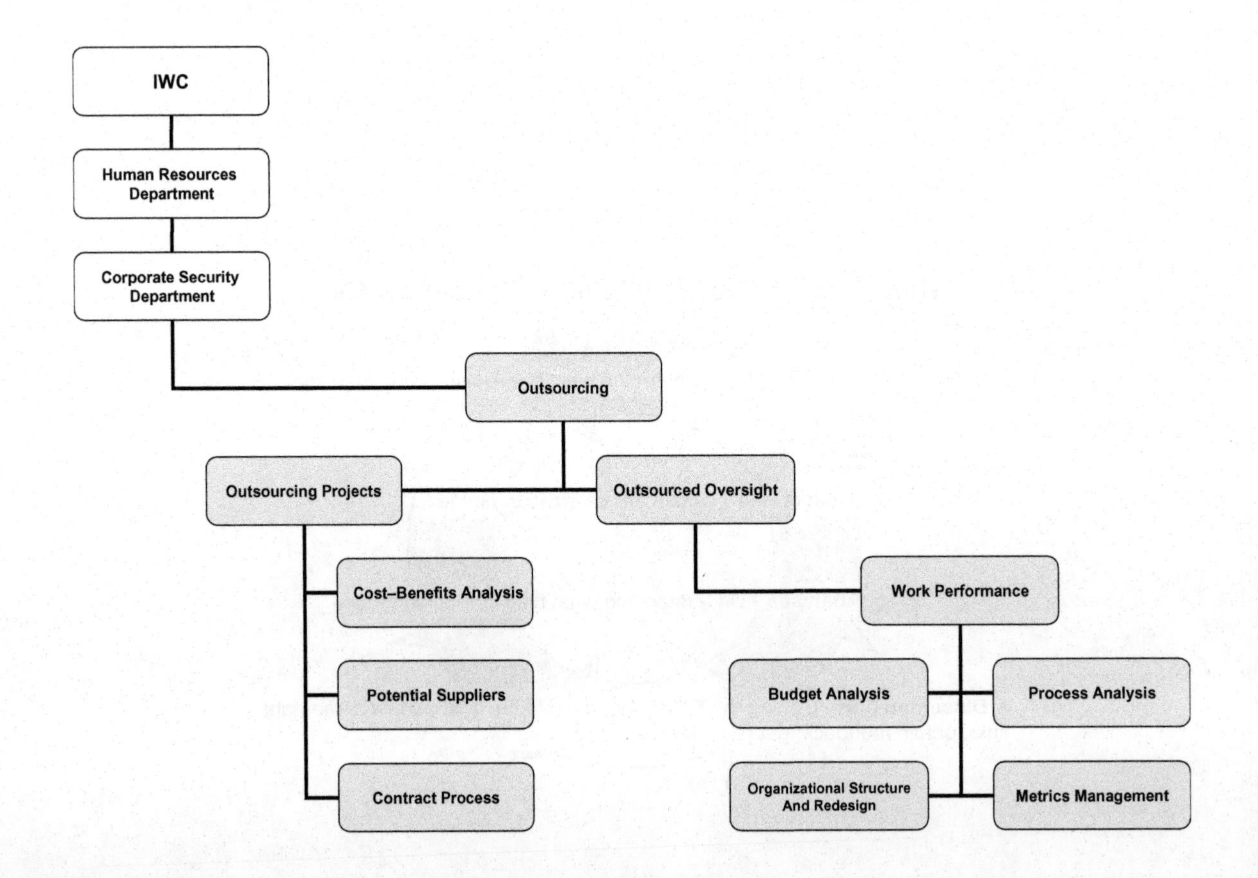
IWC
Human Resources Department
Corporate Security Department
Outsourcing
Outsourcing Projects
Outsourced Oversight
Cost–Benefits Analysis
Potential Suppliers
Contract Process
Work Performance
Budget Analysis
Process Analysis
Organizational Structure And Redesign
Metrics Management

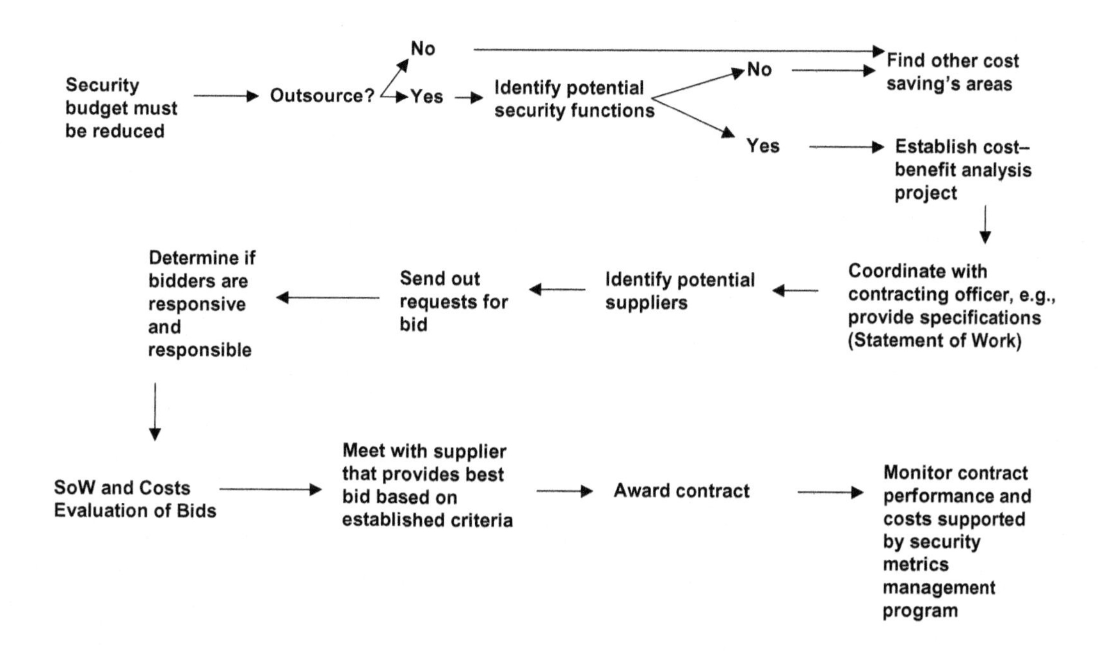
Security budget must be reduced
Outsource?
No
Yes
Identify potential security functions
No
Yes
Find other cost saving's areas
Establish cost–benefit analysis project
Coordinate with contracting officer, e.g., provide specifications (Statement of Work)
Identify potential suppliers
Send out requests for bid
Determine if bidders are responsive and responsible
SoW and Costs Evaluation of Bids
Meet with supplier that provides best bid based on established criteria
Award contract
Monitor contract performance and costs supported by security metrics management program

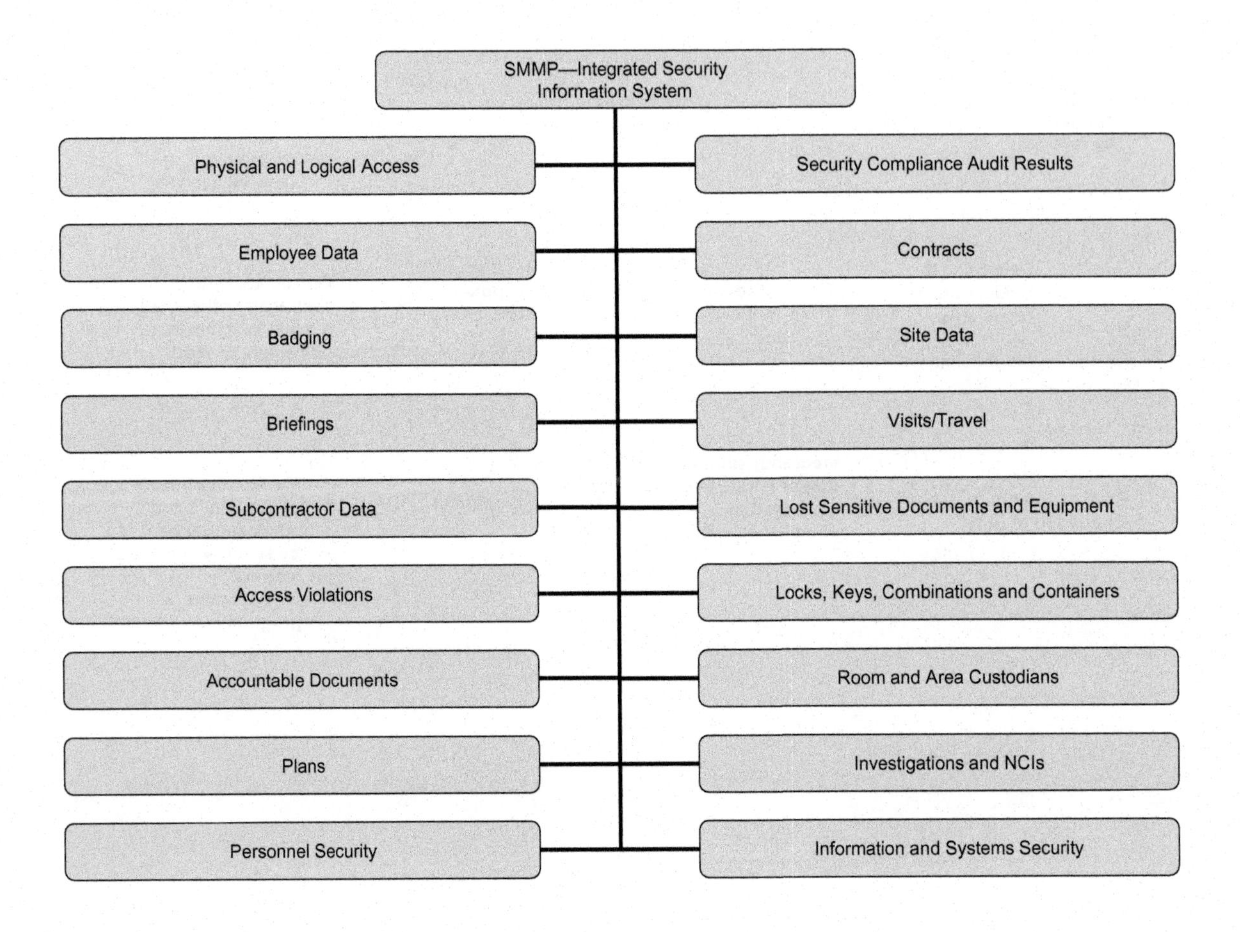
SMMP—Integrated Security Information System
Physical and Logical Access
Employee Data
Badging
Briefings
Subcontractor Data
Access Violations
Accountable Documents
Plans
Personnel Security
Security Compliance Audit Results
Contracts
Site Data
Visits/Travel
Lost Sensitive Documents and Equipment
Locks, Keys, Combinations and Containers
Room and Area Custodians
Investigations and NCIs
Information and Systems Security

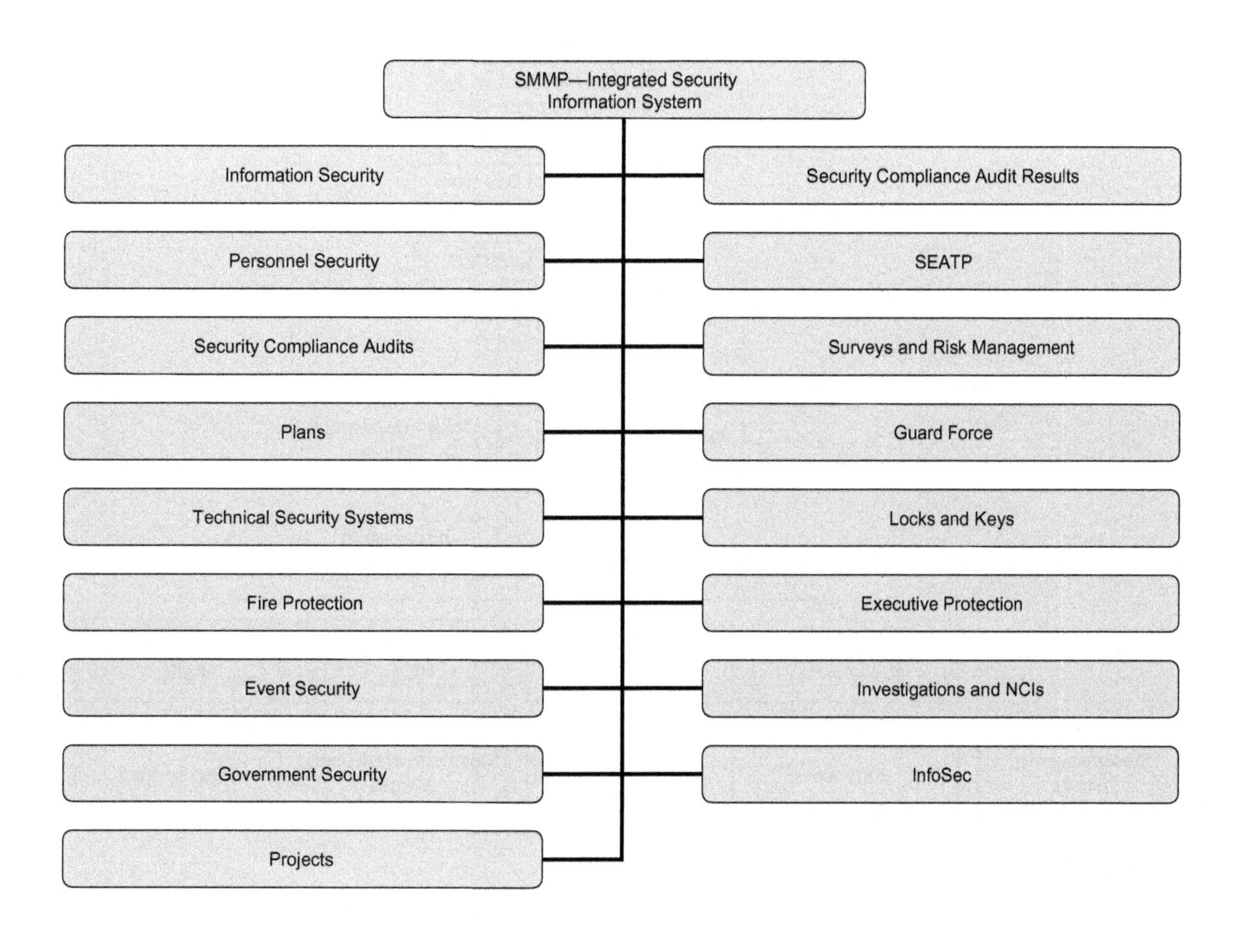
SMMP—Integrated Security Information System
Information Security
Personnel Security
Security Compliance Audits
Plans
Technical Security Systems
Fire Protection
Event Security
Government Security
Projects
Security Compliance Audit Results
SEATP
Surveys and Risk Management
Guard Force
Locks and Keys
Executive Protection
Investigations and NCIs
InfoSec

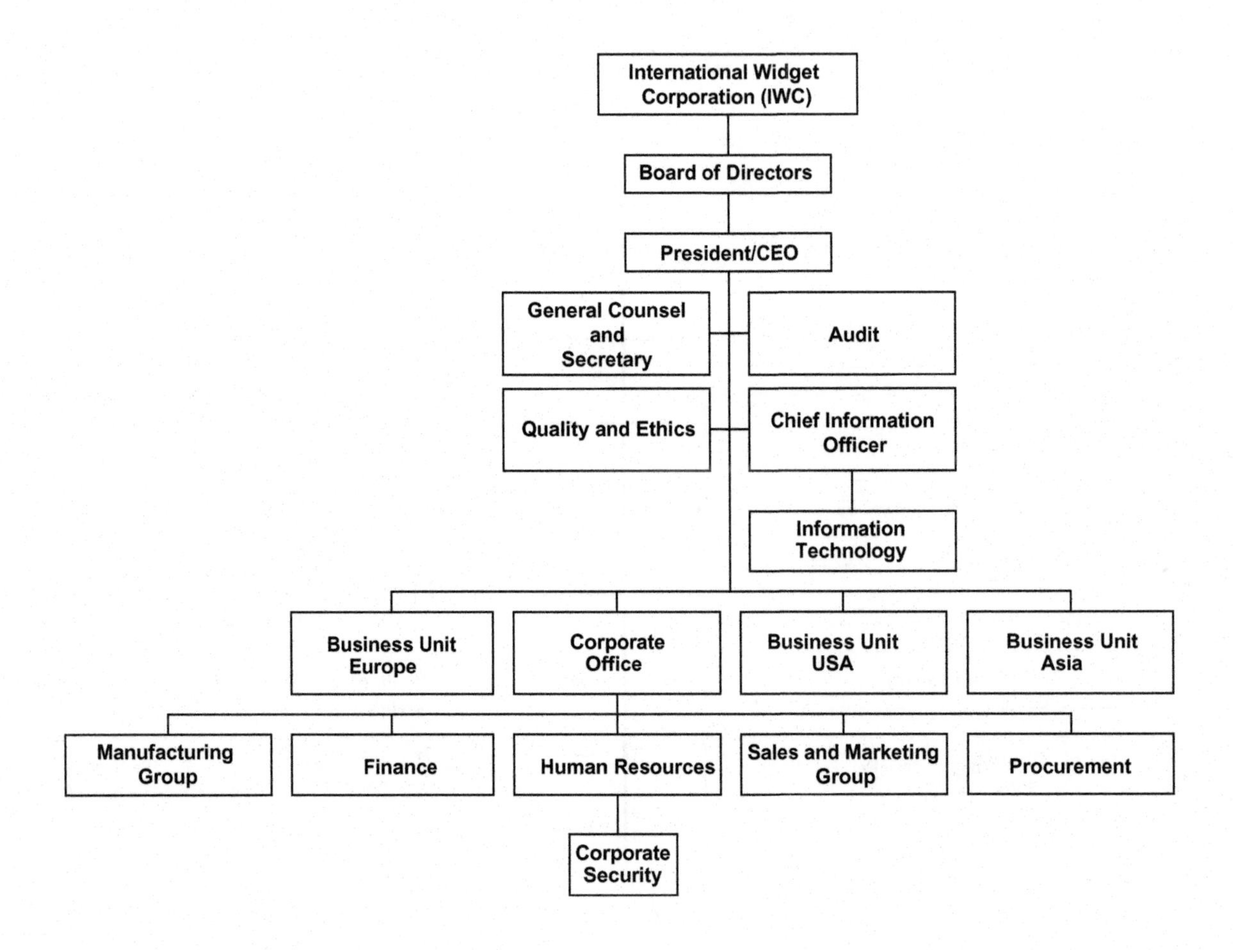
International Widget Corporation (IWC)
Board of Directors
President/CEO
General Counsel and Secretary
Audit
Quality and Ethics
Chief Information Officer
Information Technology
Business Unit Europe
Corporate Office
Business Unit USA
Business Unit Asia
Manufacturing Group
Finance
Human Resources
Sales and Marketing Group
Procurement
Corporate Security

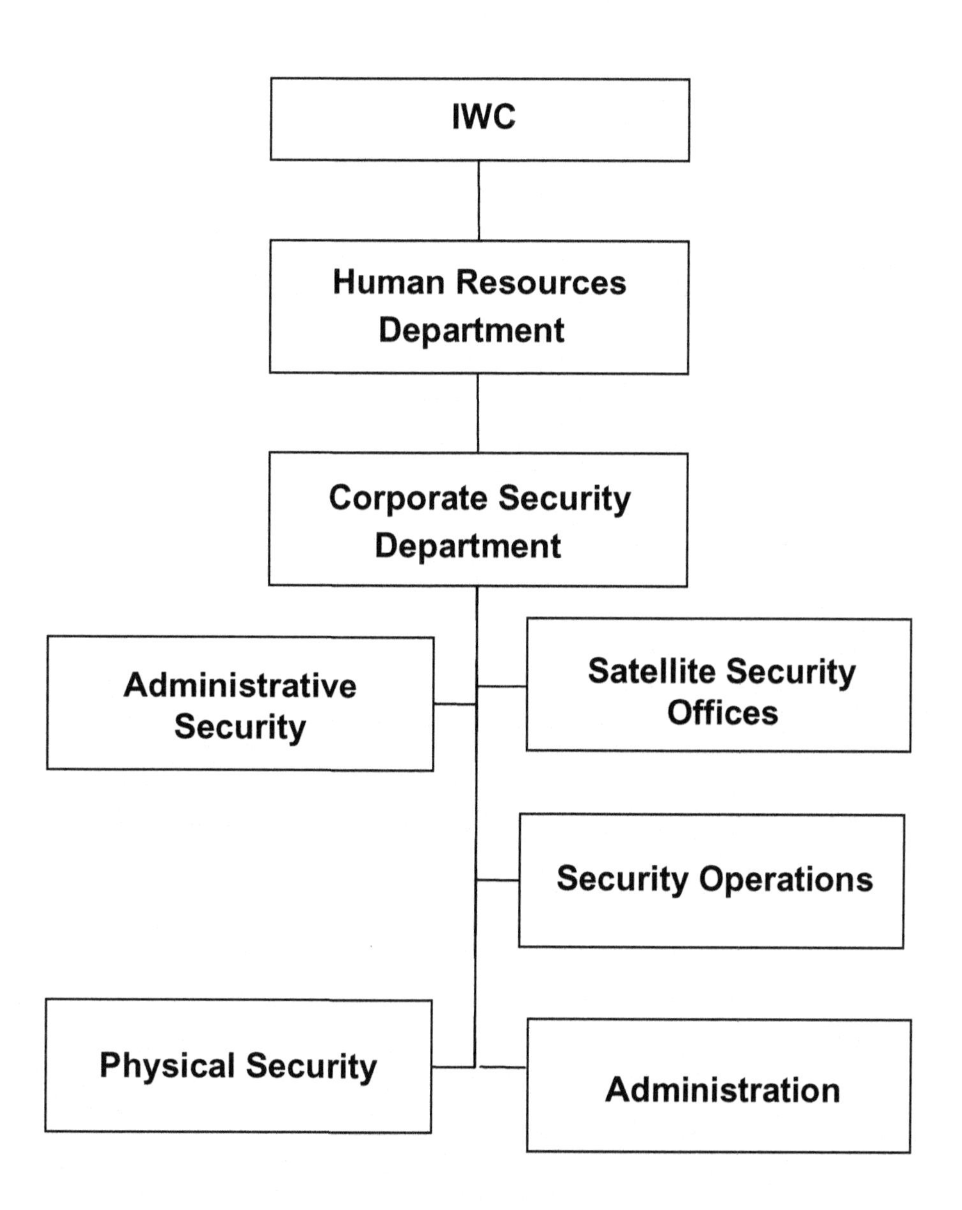
IWC
Human Resources Department
Corporate Security Department
Administrative Security
Satellite Security Offices
Security Operations
Physical Security
Administration

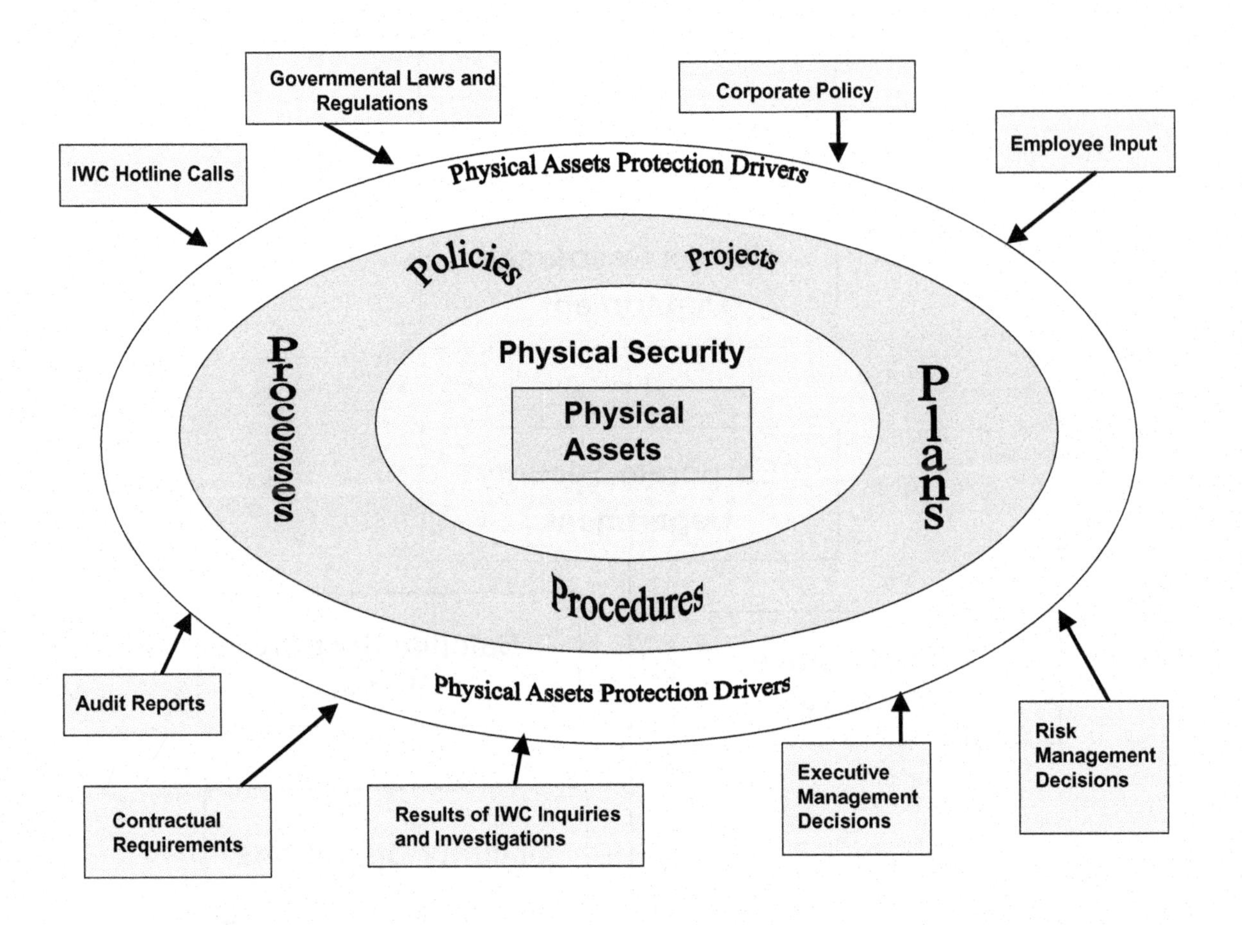
Governmental Laws and Regulations
Corporate Policy
Employee Input
IWC Hotline Calls
Physical Assets Protection Drivers
Policies
Projects
Processes
Physical Security
Physical Assets
Plans
Procedures
Physical Assets Protection Drivers
Audit Reports
Contractual Requirements
Results of IWC Inquiries and Investigations
Executive Management Decisions
Risk Management Decisions

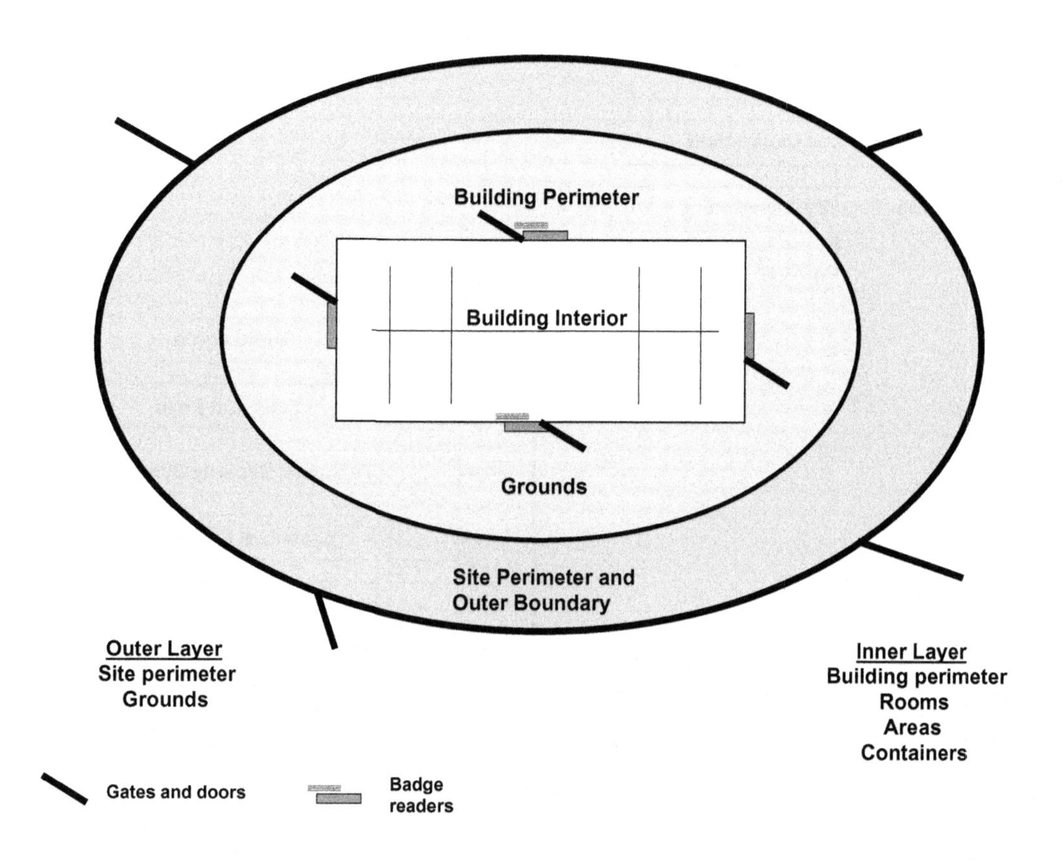
Building Perimeter
Building Interior
Grounds
Site Perimeter and
Outer Boundary
Outer Layer
Site perimeter
Grounds
Inner Layer
Building perimeter
Rooms
Areas
Containers
Gates and doors
Badge
readers

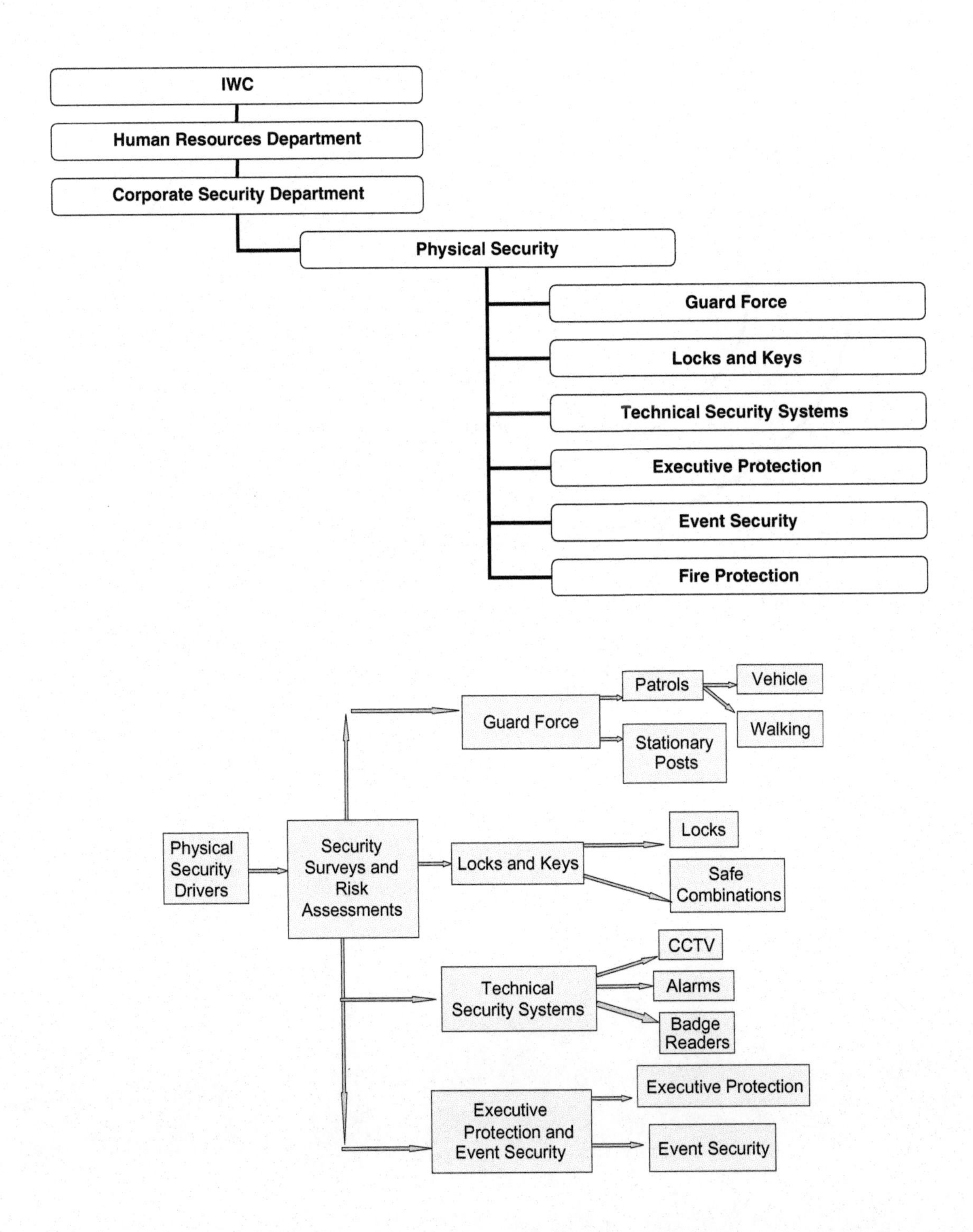
IWC
Human Resources Department
Corporate Security Department
Physical Security
Guard Force
Locks and Keys
Technical Security Systems
Executive Protection
Event Security
Fire Protection
Physical Security Drivers
Security Surveys and Risk Assessments
Guard Force
Patrols
Vehicle
Walking
Stationary Posts
Locks and Keys
Locks
Safe Combinations
Technical Security Systems
CCTV
Alarms
Badge Readers
Executive Protection and Event Security
Executive Protection
Event Security

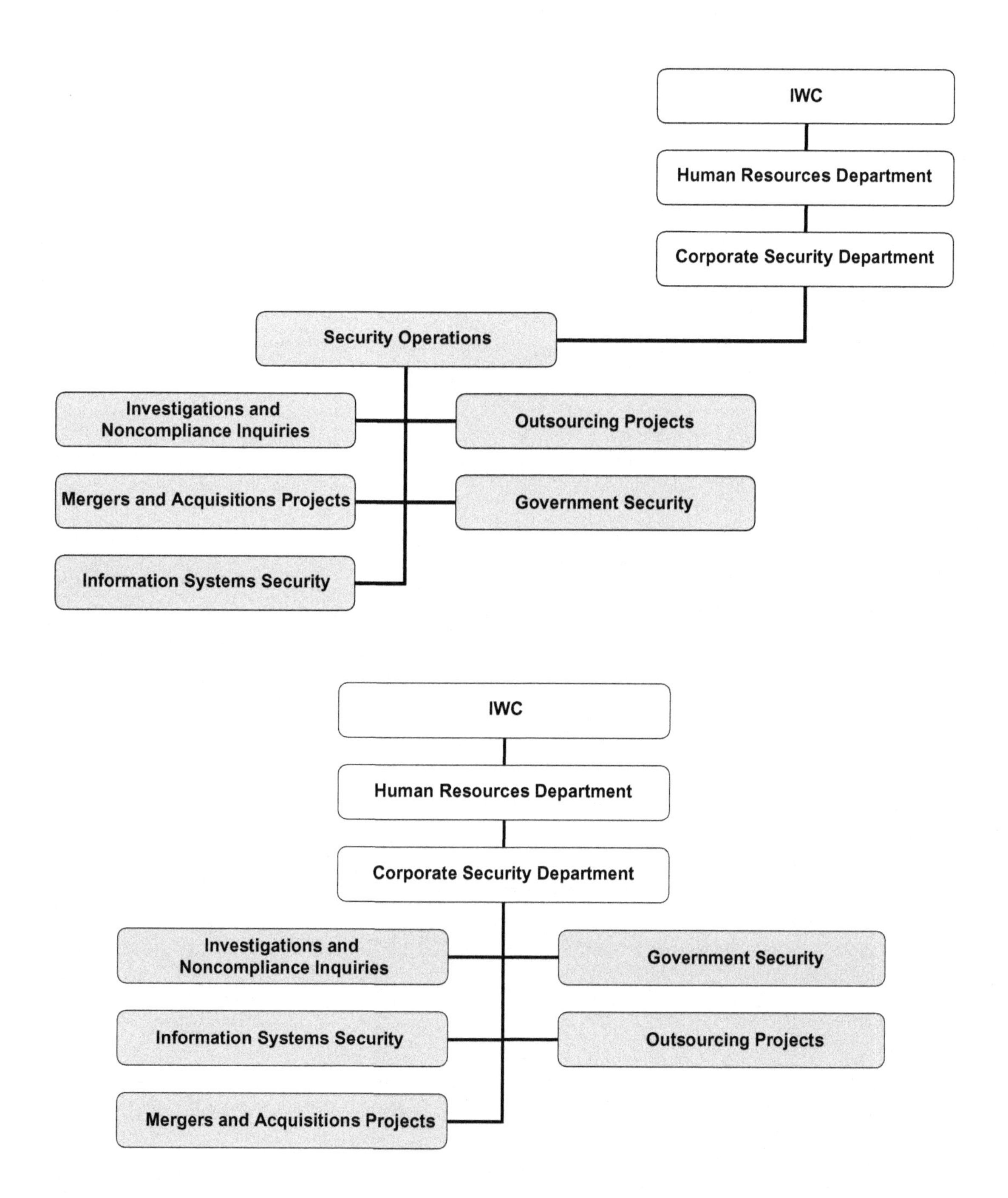
IWC
Human Resources Department
Corporate Security Department
Security Operations
Investigations and Noncompliance Inquiries
Outsourcing Projects
Mergers and Acquisitions Projects
Government Security
Information Systems Security
IWC
Human Resources Department
Corporate Security Department
Investigations and Noncompliance Inquiries
Government Security
Information Systems Security
Outsourcing Projects
Mergers and Acquisitions Projects

Index

J

K

L

M

N

O

P

R

S

T

V

W

Printed in the United States
By Bookmasters